AF352672

Energizing Neoliberalism

Energizing Neoliberalism

The 1970s Energy Crisis and the Making of Modern America

Caleb Wellum

Johns Hopkins University Press

Baltimore

Johns Hopkins University Press
2715 North Charles Street
Baltimore, Maryland 21218
www.press.jhu.edu

Library of Congress Cataloging-in-Publication Data

Names: Wellum, Caleb, 1985– author.
Title: Energizing neoliberalism : the 1970s energy crisis and the making of
    modern America / Caleb Wellum.
Description: Baltimore : Johns Hopkins University Press, 2023. | Series:
    Energy humanities | Includes bibliographical references and index.
Identifiers: LCCN 2022052799 | ISBN 9781421447186 (hardcover) |
    ISBN 9781421447193 (ebook)
Subjects: LCSH: Energy policy—United States—History—20th century. |
    Petroleum reserves—Political aspects—United States—History—20th
    century. | Neoliberalism—United States.
Classification: LCC HD9502.U52 W448 2023 | DDC 333.790973—dc23
    /eng/20221104
LC record available at https://lccn.loc.gov/2022052799

A catalog record for this book is available from the British Library.

*Special discounts are available for bulk purchases of this book. For more
information, please contact Special Sales at specialsales@jh.edu.*

Contents

Acknowledgments

I have incurred many debts in the writing of this book. I could not have done this work without the support of many mentors, colleagues, friends, and family.

Generous financial support came from many sources, including the Social Sciences and Humanities Research Council of Canada; the Transitions in Energy, Culture, and Society project of Future Energy Systems at the University of Alberta; the Department of History at the University of Toronto; the Department of Historical Studies at the University of Toronto Mississauga; the Ontario Ministry of Training, Colleges and Universities; the Centre for the Study of the United States at the Munk School of Global Affairs; and the Gerald R. Ford Presidential Library.

Several archivists and librarians helped me to track down documents and obscure books about energy. William McNitt and Elizabeth Druga of the Gerald Ford Library and Keith Shuler of the Jimmy Carter Library provided invaluable assistance to an archival novice. Kay Peterson of the Smithsonian Archives Center and Margaret Hogan at the Rockefeller Archive also introduced me to material that shaped my thinking on this project. A significant portion of my research relied on the staff of the interlibrary loan office at Robarts Library.

This book began as a dissertation at the University of Toronto, where I benefited from an outstanding and supportive dissertation committee. Elspeth Brown, Michelle Murphy, and Steve Penfold were consistently collegial, generous, patient, and brilliant. I particularly want to thank Elspeth for her unwavering support as a supervisor, mentor, and colleague. Elspeth always advocated for me and challenged me to do my best work. Her expert guidance made my work better and her ongoing support made it possible.

I am fortunate to be working with Johns Hopkins University Press to bring my first book to fruition. It has been a pleasure to work with my editor, Matt McAdam, and the rest of the staff at Hopkins Press, especially Adriahna Conway. Elizabeth Farry provided excellent copyediting support. The book's anonymous reviewers, as well as reviewers for articles with *Environmental History* and *Enterprise and*

Society and several book chapters, sharpened my thinking and led to significant improvements.

Portions of chapter 3 appeared in an earlier form in "'A Vibrant National Preoccupation': Embracing an Energy Conservation Ethic in the 1970s," *Environmental History* 25, no. 1 (January 2020): 85–109. A previous version of some of chapter 4 appeared in "'Keep Movin': *Convoy* (1978), Car Films, and Petro-Populism in the 1970s," in *American Energy Cinema*, edited by Robert Lifset, Raechel Lutz, and Sarah Stanford-McIntyre (West Virginia University Press, 2023). Portions of chapter 5 appeared in a previous form in "Energizing Finance: The Energy Crisis, Oil Futures, and Neoliberal Narratives," *Enterprise and Society* 21, no. 1 (March 2020): 2–37.

Thanks to my many teachers, mentors, and colleagues in the Departments of History at McMaster University and the University of Toronto, the Petrocultures Research Group, the After Oil School, and beyond. Thanks as well to Mark Simpson at the University of Alberta for intellectual and financial support, encouragement, and collaboration.

To the amazing Imre Szeman, my deep thanks for your mentorship, support, engagement, collaboration, and friendship. Like so many others, I have benefited and learned so much from the work that you do and the way you do it. I can't tell you how much your support has meant to me, and I look forward to future projects.

To Michael Clemens and Dave Szostak, thanks for your close and sustaining friendship over the years: the debates, expeditions, conversations, projects, and everything else.

I owe so much to the love and support of my family. Thank you to my parents, Kirk and Debbie, for instilling a love of thinking and reading in me and for encouraging and supporting me on this path. My siblings, Brittany (and Jon), Seth, and Javan, as well as Ruby, Kirk, Georgina, and Alex, make coming home so happy and fun. Finally, thank you to the Corazzas—Marco, Theresa, Dan, Elizabeth, and Sandra—for welcoming me so warmly into their family.

I could not have done this work without the love and constancy of my wife, Stephanie Corazza. Steph read every word of this book multiple times, as well as presentations, lectures, and applications. She has repeatedly and selflessly offered her time, insights, energy, companionship, and care throughout this long, arduous, and sometimes isolating process, all while completing her own PhD and starting her career. I feel privileged to have shared life with her in so many different places and circumstances over the past decade. She enriches my life every day.

This book is dedicated to her.

Energizing Neoliberalism

Energy in Crisis

Crisis partakes of a metaphysics of history.　　　　　—JANET ROITMAN

The 1970s energy crisis never actually happened, but it mattered.[1] Of course, many Americans imagined themselves beset by an unprecedented crisis of energy in the winter of 1973–74, but they disagreed on its nature. Some blamed government policy for fostering gluttonous oil consumption or anemic production. Others viewed lines for gasoline, persistent inflation, and forecasts of declining oil supplies as portents of the limits to growth. Populists blamed the greed of OPEC (Organization of the Petroleum Exporting Countries) or "Big Oil" while pundits criticized foreign policy. The crisis usually seemed to be rooted in resource scarcity, but at other times it seemed to be about resource prices. Public discourse often focused on oil, but it also covered natural gas, electricity from various sources, nuclear power, and the possibility of a transition to alternative energy sources like solar or wind. In the end, the world did not "run out" of energy. More oil was discovered, contributing to gluts by the mid-1980s, and there never was any danger of running out of energy as long as the sun shone. Given such contradictions and confusions, it is difficult to maintain that in the 1970s the United States endured a coherent energy crisis with a clear cause, beginning, and end. Even so, the events and debates that were assembled into an energy crisis reshaped postwar American history and remain deeply relevant for contemporary sustainability struggles. It is the goal of this book to explain how the 1970s energy crisis came to matter and, in so doing, to reflect on the centrality of crisis to life in late modernity.

The energy crisis is often recalled as the domestic effect of a geopolitical crisis that began in the Middle East on October 6, 1973. On that day, an Arab coalition led by Egypt and Syria launched a surprise attack on Israel to reclaim land lost during the Six-Day War of 1967. The attack was a surprise because it fell on Yom Kippur, Judaism's holiest day of fasting and rest from labor. The brief war took a heavy toll. Intending to isolate its adversary, the Organization of Arab Petroleum Exporting Countries (OAPEC) declared that it would cease oil shipments to any nation that supported Israel in the conflict. When the United States airlifted military aid to Israel to counter Soviet support for the Arab coalition, OAPEC cut oil production by 5 percent per month and vowed to continue to do so for each month that Israel remained in the occupied territories. Saudi Arabia announced a 20 percent cut, flexing its muscles as the world's most powerful oil-exporting nation. Next, OAPEC embargoed all oil shipments to the United States and the Netherlands.[2] Anxiety and anger ensued from many American citizens as the United States faced tightening supplies and rising prices.

The Yom Kippur War ended after only twenty days, but the oil embargo lasted until March 1974, and OAPEC's price increases haunted US social and political life into the 1980s. Nearly 17 percent of US oil imports, which accounted for almost a quarter of the country's oil consumption in 1973, came from OAPEC nations, so the use of the "oil weapon" had sudden, if short-lived, effects on the supply of America's lifeblood commodity. The price of oil spiked from just below $3/barrel in September 1973 to nearly $12/barrel in December.[3] The situation worsened, as did the perception of a crisis, as panic buying led to long lines at gas stations, trucker protests slowed highway traffic, and analysts forecasted dire futures. Governments tried to respond in a variety of ways, including price controls, gasoline rationing, diplomacy, and entreaties to a bewildered public to find ways to save energy.[4] After the embargo, oil prices stabilized but remained above their pre-embargo levels; talk of an ongoing energy crisis continued to circulate.

An acute sense of crisis returned in 1979, when another wave of shortages, price spikes, and protests hit the United States in the wake of the Iranian Revolution. Strikes in the Iranian oil sector and the turmoil of revolution cost the United States hundreds of thousands of barrels of imported Iranian oil.[5] Iran was the second-largest supplier of imported oil to the United States, and the loss of its oil translated into an 8 percent reduction in US oil imports and a 150 percent price increase. Although Saudi Arabia agreed to compensate by increasing production, prices shot up as the still-vivid memory of 1973 drove many Americans to respond again with panic. Oil companies and consumers purchased more oil more frequently than they otherwise would have, driving prices from \$13/barrel to \$34/barrel by 1981.[6]

Although these "oil shocks," as they were called, played an important role in triggering claims of an energy crisis, they do not tell the whole story. Focusing on them, as do many accounts of the energy crisis, implies that the crisis was the self-evident outcome of supply and demand dynamics within larger geopolitical histories of oil.[7] The trouble is that the energy crisis was *not* a simple, self-evident matter of energy supply and demand, as endemic disagreement over its nature suggests. As I argue in this book, the energy crisis was a complex assemblage of energy-consuming practices and narratives of crisis that were deeply cultural and hotly contested. In order to understand the full import of the 1970s energy crisis it is vital to examine its relationship to these competing narratives, meanings, attitudes, and values.

A Cultural History of the Energy Crisis

This book is primarily about the fundamental discourses and narratives that sought to make meaning out of the energy crisis, rather than the geopolitical and policy dimensions of the crisis that occupy other accounts. This attention to discourse is rooted in the materiality of fossil fuels.[8] Unlike wood resources for earlier generations, oil, for instance, is largely invisible to most users. We interact with oil strictly as consumers at the end of long supply chains composed of remote extraction sites, guarded refineries, buried pipelines, and sealed-off gasoline tanks, or in the form of plastic products no longer easily recognized as oil. The same is true for the fossil fuels that most Americans experience as electricity or home heating—they are known through the electrical outlet or the furnace. Most people necessarily rely on expert claims, forecasts, and prices, as well as popular narratives and other forms of discourse, to situate their experience of energy abundance and scarcity. Discerning a crisis of fossil fuel energy is not as straightforward as walking through a denuded landscape. Such a crisis must be put together and then put to work discursively.

Public discourse in the 1970s put the energy crisis to work in a variety of ways. US political leaders framed it in the culturally significant terms of "independence." President Richard M. Nixon named his keystone energy policy "Project Independence," and in public statements on energy he recited the various ways in which quotidian twentieth-century freedoms required cheap and abundant energy. Most Americans consumed fossil fuels to hold down a job, spend time with family, attend religious worship, participate in popular culture, and heat their homes.[9] The functional independence of US citizens, Nixon implied, relied on the provision of abundant and affordable energy resources. Such a notion of independence as presupposing far-flung energy systems was a long way away from the Jeffersonian yeoman, and it would have sounded strange to the authors of the Declaration of Independence, who had in mind political freedom from external coercion or control. It reflected material changes in American life as well as a narrative construction of the energy crisis as a threat to America's founding ideals and culture. And it is just one example of the many ways in which the energy crisis, and the fossil-fueled capacities that it undermined, worked as cultural symbols in wide-ranging debates about freedom, capitalism, nature, nation, and the self. Examining the energy crisis as cultural history highlights the various and contested meanings of energy and the crisis in the 1970s.

Key to my cultural analysis of the energy crisis is developing a clearer view of the historical development of a petroculture in the United States. Petroculture is a recent neologism that names the deep connection between energy resources and the culture and politics of a given society. It describes a society that depends on fossil fuels and is, in turn, contoured by that dependence. Its built environment, material practices, political concerns, and institutions, as well as a whole range of habits, values, and beliefs, are intertwined with the capacities afforded by fossil fuels and the demands that their provision makes on a society and its members.[10] This concept expands the possibilities for studying energy resources from their relationship to industrial production, trade, and pollution to the realm of thought and cultural practice, including meaning, identity, and ideals.

Although it privileges the social influence of oil, petroculture does not preclude the transformative impact of other fossil fuels. Coal and natural gas share key features with oil and foster similar forms of abstracted mass energy consumption that make up petrocultures. Even electricity, which is a form of energy that has marked American modernity at least as starkly as has oil, can be analyzed in terms of petroculture at certain historical moments. It is worth recalling that for much of the twentieth century, most electricity in the United States was generated by burning coal, natural gas, and oil.[11] At those moments, electricity existed within the logic of centralized systems of energy provision and mass consumption common to all fossil fuels.[12] Moreover, public discourse about the 1970s energy crisis often conflated oil with other forms of energies, bespeaking the unique cultural power of oil as a source of endless abundance.

Historians have long studied the history of energy resources, but they have yet to consider petroculture as an analytic concept.[13] Some have used the concept of an "energy regime" or "energy system" to explore the broad impact of energy resources, but their analyses tend to shy away from representation and meaning in favor of political economy.[14] In recent years, however, the analysis of petroculture has flourished in literary studies as part of a new subfield called the energy humanities. It has sparked interest in the presence of energy in literature and various other forms of representation, and scholars have toyed with the idea of literary periodization according to energy, but it has yet to shape sufficiently the stories that we tell about the culture and politics of the past.[15] There is still much that we do not know about the historical dynamics of petrocultures.[16]

Nevertheless, recent work in energy history has produced important insights into the roots of US fossil fuel dependence. Some note the year 1885 as a milestone in this process because it was in this year that the United States began to derive more energy from fossil fuels than from renewable resources, creating an imbalance that has only grown over the ensuing decades.[17] Christopher Jones points to infrastructure, arguing that by 1930 a century of building new energy-delivery systems incubated US modernity and oriented the nation toward extensive fossil fuel dependence.[18] Matthew T. Huber, in contrast, points to the New Deal as a key moment in US energy history insofar as the socioeconomic transformation that it represented "centered on the mass consumption of energy."[19] In their efforts to rescue capitalism from the Great Depression, New Deal programs set the stage for high wages and private home ownership while creating a stable and sufficiently affordable market for oil. These accounts expose vital roots of the postwar explosion of energy consumption that fostered an energy crisis in the 1970s. But postwar petroculture was not merely the natural outgrowth of these infrastructural and ideological antecedents, as important as they were. There is more to the story.

This book builds on these accounts of fossil fuel dependence by turning to the postwar period as a key moment of petrocultural formation in the United States. Although energy infrastructures and the (sub)urban modernity that they fostered were necessary antecedents, US petroculture actually emerged in the postwar period in parallel with the world's "great acceleration" in global energy use and CO_2 emissions.[20] The energies that fueled the processes of concentration and mass politics traced by Jones began to fuel decentralization and an increasingly privatized individualism through an elaborate energy-consuming culture designed to ward off the threat of communism. The growth of this mass consumption culture and its concomitant concepts, values, and anxieties, I argue, was the primary cause of the 1970s energy crisis that in turn funneled US political culture toward ever-greater individualism and reliance on free markets. This book explains that process by positing the Cold War politics of capitalism as a driving force in the formation of US petroculture that expanded the New Deal vision of American life materially and ideologically. The desire to differentiate American capitalism from Soviet communism intensified the mass energy consumption presupposed by the New Deal vision of American life. It also fostered crisis narratives about energy, the environment, and mass destruction that helped to generate a crisis of energy in the 1970s and fostered neoliberal forms of political culture.

Turning to the Right

The quarter century after the Second World War was a distinctive period in American history that is often recalled as a time when the nation was on a path to greater social and economic equality and security for its citizens. Dubbed the "Age of Keynes" by one journalist, it was a period of broad commitment to implement and extend the ideas of the British economist John Maynard Keynes regarding the proper role of government intervention in social and economic matters, as well as the nature and role of individual liberty.[21] Although the idea of a "liberal consensus" has been roundly and rightly criticized—no nation ever achieves complete unity—there existed a broad and effectual commitment to a welfare state and regulated markets in postwar America.[22] Many leaders and experts believed that government intervention could resolve social and economic problems, as evidenced by President Lyndon B. Johnson's Great Society programs. Buoyed by the pain of the Great Depression, Keynes's ideas validated deficit spending and thorough government regulation to counteract the tendency of capitalism toward inequality and cycles of boom and bust. Amid rising prosperity after years of depression and war, argues Thomas Borstelmann, relying on state expertise to manage "a modern economy came to seem like common sense."[23] Why the United States veered so sharply from this path is a question that has occupied historians ever since.

Historian Jefferson Cowie offers an explanation that frames the period from 1930 to 1970 as a "great exception" to the warp and woof of American history, a time when American liberalism prioritized "collective economic rights" rather than the interests of elites.[24] The labor movement secured for many non-elite Americans an unprecedented share of national wealth while Keynesianism guided economic policy. "As a result," writes Cowie, "more income, more equality, more optimism, more leisure, more consumer goods, more travel, more entertainment, more expansive homes, and more education were all available in the postwar years to regular people than at any other time in world history."[25] Why couldn't it last? Cowie insists that the seeds of the demise of the great exception were present at its birth. New Deal or Keynesian liberalism existed in tension with the cultural inheritance of Jeffersonian individualism and never resolved long-standing fissures of immigration, race, religion, and a range of other issues.[26] When these fissures widened in the 1970s, New Deal liberalism broke apart, a turn of events that Cowie describes as "quite understandable, perhaps even axiomatic."[27]

By highlighting the pivotal role of energy and the energy crisis to the course of postwar US history, this volume complicates narratives like these, which tend to treat this period as a model for the future, if only the resurgent political forces of conservatism that ended it could be overcome. Cowie, for instance, claims that politics rather than economic structure drove the great exception, implying that returning to the right politics can turn the exception into the rule.[28] His argument echoes the sentiments of the Marxist literary critic Raymond Williams, who claimed in 1958 that the principal task of the future would be to "use . . . our new resources to make a good common culture; the means to a good, abundant economy we already understand."[29] Although politics were certainly important, these positions stand in need of revision as we learn more about the deep links between energy resources, politics, and culture. It is not clear that anyone understands how to create an abundant economy as Cowie describes it without substantial fossil fuel inputs. The great exception and its Keynesian economy benefitted from the development of an elaborate petroculture in the United States. The socioeconomic trends in union membership and share of annual income that constitute the great exception coincide with levels of fossil fuel consumption that cannot be a feature of any sustainable future. And, like the great exception, US petroculture contained within it the seeds of an energy crisis that would help to shift the nation's political culture in a rightward direction.

The dominant historical paradigm concerning the decline of the great exception tends to focus on the rise of neoconservatism. Historians have analyzed conservative ideas and social movements in the context of anticommunism, social changes stemming from sixties social movements, and religion, among other factors.[30] Several scholars working on the social and political history of the New Right in the second half of the twentieth century stress the centrality of racial politics and white racial resentment to its ascendency. They privilege white efforts to maintain de facto segregation and disparities of wealth and social power in the wake of the sixties civil rights movement while adopting an evolving language of colorblindness and commonality. The New Right, according to these works, is primarily a white and neoconservative political movement formed in opposition to African American civil rights and social integration.[31] The postwar suburban American dream that the energy crisis threatened was primarily a white reality by design, defended by conservative forces.[32]

Paying attention to the cultural politics of the energy crisis ought to shift our attention from neoconservatism to neoliberalism, which differs from the former in the depth of its commitment to individualism and market rationality. As Matthew Lassiter notes, much of the analysis of the New Right is wedded to the "red-blue binaries of journalism and punditry," tying racial politics and free market policies largely to the conservative, Republican side of the partisan divide as explanations for the decline of Keynesian liberalism.[33] Energy crisis discourse blurs this binary and suggests that the story extends well beyond neoconservatism, even if, as Meg Jacobs has shown, political conservatives capitalized on the energy crisis.[34] The political history of the twentieth century cannot be reduced to party politics. How should we understand a writer like Wendell Berry, an anti-capitalist conservative who advocated community, tradition, and stability over market disruption; Edward Kennedy, a socially liberal Democrat who supported oil futures financialization and a politics of oil abundance; or Denis Hayes, an Earth Day organizer who supported free market oil pricing? Their engagements with the energy crisis and views on the market do not align neatly with the dominant binary. Paying greater attention to the role of neoliberalism better accounts for the thoroughgoing economization of US political culture beyond the red-blue and conservative-progressive divides.[35] The ascendency of neoliberalism is not reducible to a New Right narrative, even as the world that the energy crisis threatened was largely the preserve of middle-class white Americans.[36] Fossil fuels undergirded the great exception, and the crisis that followed created the conditions in which a deep and thoroughgoing market epistemology could flourish.

Defining neoliberalism is a difficult task, in part because, as Dieter Plehwe notes, it is "anything but a succinct, clearly defined political philosophy."[37] Neoliberalism has multiple points of origin, advocates, and affiliated institutions, and it is often mistaken for nineteenth-century liberalism and classical economic theory. But it is possible to delineate its key features, three of which are important for the argument of this book. First, neoliberalism is a social and political ideology that privileges free markets and individual liberty understood primarily in economic terms. Neoliberals advocate free enterprise, competition, and the unparalleled value of the price mechanism and posit state cultivation and protection of free markets because they recognize that capitalist markets are fragile technologies that must be built and maintained; they do not exist in a state of nature.[38] Second, neoliberalism is a totalizing epistemology that sees the world through the lens of market competition. It extends economic logic to areas of social and political life once considered to be beyond the purview of economics, generalizing entrepreneurialism throughout society and positing the market as an efficient information processor from which truth emerges.[39] Finally, neoliberalism is a "governing rationality" that uses market discipline to impose austerity, which was a point of overlap with mid-century environmentalism.[40] Historically, it has served the interests of US and Western hegemony by extending market relations around the world through institutions like the World Bank and International Monetary Fund. It is a class project within the United States and an imperialist project when extended abroad.[41]

The propagation of neoliberal perspectives and ideas during the 1970s occurred at many different levels of discourse. Typically, studies of neoliberalism consider only one or the other. Either scholars write about the intellectuals and institutions responsible for neoliberal theorizing, such as Friedrich Hayek, Gary Becker, the Chicago School of Economics, and the Geneva School, or they attend to the cultural politics of neoliberalism.[42] The energy crisis affords an opportunity to consider the contribution of unique forms of expert and popular discourse to the rise of a neoliberal political culture. Thus, rather than focus on economic theorizing or the ideological contours of suburban life, this book considers how experts addressing the energy crisis engaged in debates that assembled a crisis of energy amenable to neoliberal intervention while at the same time "petro-populism" flourished in popular culture. Examining the expert and the popular together offers a fuller picture of how the struggle over the meaning of the energy crisis at multiple levels fostered a neoliberal orientation toward life.

By connecting US petroculture, the energy crisis, and neoliberalism, I develop two further senses of neoliberalism. In the first instance, neoliberalism tends toward anticipatory temporalities, as in Melinda Cooper's claim that it "installs speculation at the very core of production."[43] By normalizing speculation as a ubiquitous practice, most saliently in metastasizing derivatives markets, neoliberalism fosters anticipation as a way of being in the world that orients its subjects toward the future. It is a regime of anticipation that produces and uses crises, which are inherently speculative, as a way to generate value beyond any conceivable limits to growth.[44] For Cooper, the 1970s oil crises accelerated this shift toward neoliberal speculation (and then anticipation) by manifesting the earth's limits to growth through oil scarcity and forcing US capitalism to turn to other, nonmaterial avenues to generate value no longer possible under conditions of scarcity, such as the bio-economy.[45] But this book demonstrates that neoliberals also celebrated the power of speculative markets to revive the resource economy. So, while environmentalists speculated widely about apocalyptic and utopian energy futures, neoliberalism adapted the injunction to speculate to its rhetoric about the possibility of unlimited abundance through free markets.

In the second instance, neoliberalism is cultural. It is a structure of feeling that informs the experience of meaning and value.[46] Neoliberal culture exerts pervasive pressure to extend market rationality to every area of human life, to disavow any robust sense of "society," and to imagine oneself as an entrepreneur-consumer.[47] The historical development of this structure of feeling owes significant debts to fossil fuels. As Huber notes, the consumption of fossil fuels enabled the development of highly privatized selves who could "imagine themselves as masters of their own life severed from ties to society and public forms of collective life."[48] Building on this argument, I show how fossil fuels also fostered a lived experience of acceleration, the anticipation of future abundance, and the celebration of risk that the energy crisis laid bare. The Cold War politics of capitalism and its attendant petroculture established the conditions for this neoliberal structure of feeling, though the tipping point occurred in discourses and experiences of a *crisis* in US petroculture.

Questioning Crisis

Modern history, to paraphrase a widely misquoted aphorism, often seems like one damn crisis after another. The news media constantly declares and decries crises that politicians, CEOs, and others try either to amend or to repress. Scientists and environmentalists warn about impending crises of extinction and climate change. Activists protest inequality and discrimination said to be at crisis levels. Indeed, as I wrote this book, the global COVID-19 pandemic sparked talk of crisis in nearly every corner of society, from public health and education to small business, media, families, and individuals. Even the energy crisis, first noticed in the 1970s, is ongoing and unresolved.[49] It is therefore important to ask what is rarely asked: What, exactly, do we mean by "crisis," and what does it take to make one?

Crisis is hard to pin down. A concept with a complex history and many meanings, crisis has ancient roots in medicine, where it referred to a turning point in the progress of a disease leading either to recovery or death.[50] Crisis also has associations with judgment, decision, or critique, and in common contemporary usage it means something like a turning point in a difficult time that impels a decision, though several variations exist.[51] Like the media, scholars rely increasingly on the power and ambiguity of crisis to create meaning out of an endless stream of events. Crisis is often the starting point for historical narratives, as indeed it is in this book's examination of what came to be called *the* energy crisis. Both left and right narrate the past around crises or write histories in order to illuminate contemporary crises. World wars and financial crises shape narratives of the twentieth century, and the present climate crisis has spurred new interpretations of human history. Social scientists and critical theorists are particularly enamored by the activist demand inherent in crisis to create a different future. This demand forms the core of social science critique, but it also creates a blind spot when scholars accustomed to invoking "crisis" to justify research and the political claims that accompany it do so unreflectively. As the anthropologist Janet Roitman notes, much of critical theory fails to reflect on the close links between crisis and the primary mode of contemporary academic engagement: critique.[52] We have become accustomed to claiming crises without "problematizing crisis" itself.[53]

One of this book's starting points is that crises are not mere objects for study that happen "out there" in the world. Crises are made, and claims of crisis are deeply political. My point is not to deny the usefulness or truth value of crisis claims, but rather to think reflexively about the genesis of crisis claims and their consequences. Building on the insights of Reinhart Koselleck, Roitman insists that crisis is cultural and discursive: it emerged within modernity's demand for a difference between past and future marked by progress.[54] The ubiquity of crisis claims, she argues, is an outgrowth of the decline of transcendent metanarratives, whether religious, modernist, or otherwise. In a time when appeals to transcendence or reason no longer resonate, crisis demarcates what is "historical" from the undifferentiated stream of past events.[55] Without crisis, it can seem nearly impossible to discern change and meaning in the passage of time since older standards of significance have fallen away, been discredited, or been repressed. To claim a crisis is to assert meaning and control in a world where meaning is made rather than discovered.

The energy crisis affords an opportunity to reflect on crises as political contests over meaning and as politically flexible assemblages of discourse and action. This book shows how different practices of energy consumption, forms of energy, and discourses about energy were *assembled* and *mobilized* by individuals and institutions with different interests in the 1970s to bring an energy crisis into being.[56] Whether or not an "actual" energy crisis existed, whatever that would mean, widely circulating claims of an energy crisis were crucial to the neoliberal evolution of postwar US political culture. The energy crisis emerged in a context of proliferating crisis claims—of the environment, population, race relations, and culture—and functioned in many ways as a crisis of crises. Different commentators saw in it the causes or consequences of other crises: of rapacious capitalism, incompetent government, or declining American power, among others. It helped to demonstrate the reality of those crises or to explain them more fully.

Unlike other accounts of the energy crisis, this book shines light on the role of environmentalists and their allies in the construction of the energy crisis and in the neoliberal turn in US culture. Many environmentalists and others seeking to mobilize the energy crisis to spur a transition to a more sustainable social order posited a crisis rooted in America's economic system and its cultural values organized around growth. Yet neoliberals proved remarkably adept at embracing claims of energy crisis, sharing environmentalist support for austerity, and channeling crisis discourse in the direction of free market regulation and a neoliberal conception of individual freedom. The affect of crisis proved useful in catalyzing the evolution of US political culture in a direction that many participants in energy crisis discourse initially abhorred. Central to that affect was a series of speculative and anticipatory discourses that fueled talk of an energy crisis and fit snugly in a neoliberal view of the world of austerity through market discipline, though always with the promise of future abundance.

Crisis narratives about energy and neoliberalism in the 1970s intersected on the grounds of anticipation and speculation. Crises are inherently anticipatory, positioned as they are in demands that the future be different from the past and claims that present events *could* become worse without positive action in the present. Neoliberalism, too, thrives on speculation and anticipation in the form of financial markets and on the burden on individuals to make the right choices to ensure a better future. The point here is not to dismiss prudence or responsibility but rather to highlight the incredible pressure and responsibility neoliberalism places on individuals for their future. The assembling of an energy crisis between 1970 and 1980 grew out of and generated anxious speculation about the future of US capitalism, as well as concern for the prospects for a healthy and stable future for the world. Indeed, the energy crisis circulated predominantly as a discursive and speculative crisis about the possibility of prolonged or more extreme shortages in the future; and it generated competing visions of the future that nudged the nation closer to market fundamentalism. Present prices and scarcities were often taken to portend a worse state of affairs in the future.

Energy politics thus played a key role in the circulation of neoliberal discourses of economic freedom and the wisdom of markets. Those discourses resisted regulatory efforts to manage energy consumption and eventually turned to financial speculation to restore US control over its energy prices and supplies. The energy crisis also included environmentalist critiques of the sustainability of carbon energy resources that spawned anxieties about the precarious postindustrial future of the United States. Environmentalist critique undermined the legitimacy of Keynesian liberalism, bolstered the neoliberal argument for austerity, and perpetuated a politics of crisis first developed during the early years of the Cold War, which free market advocates adapted to their own purposes, sometimes with environmentalist support. The articulation of a "crisis" was therefore an insufficient foundation on which to build a large-scale energy transition. Crisis instead proved to be a durable and flexible construction that could support competing visions of the future.

Outline of the Book

By exploring competing energy crisis narratives, this book shows that the 1970s energy crisis initiated a key transition in the history of neoliberalism and in the cultural and political history of the postwar United States. Over the past twenty years, scholars have increasingly come to see the 1970s as a "pivotal decade" in which the energy crisis played a key part.[57] In geopolitics, the crisis demonstrated the limits of US power. Domestically, it facilitated the political success of the New Right. The oil embargo in 1973–74 fed existing fears of national decline initiated by failure in Vietnam, tilling the soil for the renewal of US nationalism in the Reagan Revolution.[58] It ignited debates and political maneuvering over energy policy in Washington, particularly regarding oil price controls, and worked as a case study in the limits of government power that aided the ascendency of New Right politicians and policies.[59] In economic policy, higher oil prices accelerated America's move from manufacturing powerhouse to a postindustrial service economy.[60] Timothy Mitchell's influential account treats the energy crisis in the larger history of democracy, noting that it both altered US relations with Middle Eastern oil states and supported the revival of "the market" as a tool of governance that fundamentally limited democratic decision making.[61] Huber, on the other hand, focuses on the relationship between energy and the cultural politics of capitalism. He argues that the spatial distribution of the population in far-flung suburbs and forms of leisure normalized by oil abundance produced a regime of consumption that conflicted with the materiality of oil to produce a crisis in the early 1970s. Popular perceptions of the energy crisis as "engineered by anti-competitive forces," he argues, amplified "an emerging neoliberal idolatry of a pure, free, and . . . *competitive* market" in need of protection from "unfair" influence.[62]

I build on these accounts while seeking to broaden and extend their arguments regarding the central importance of the energy crisis to the political culture of the United States in the critical decade of the 1970s. This book offers a richer cultural history analysis of the lead up to the energy crisis, paying particular attention to the cultural politics of Cold War capitalism, and it explores the energy crisis itself as primarily an interpretive struggle that aided the rise of neoliberalism. Using the energy crisis to historicize and explore the role of crisis and anticipation in the genealogy of neoliberalism broadens the import of the crisis beyond the rise of neoconservatism in US party politics. It provides a way of understanding the cultural and intellectual currents that popularized the free market as a solution to myriad social and political problems. Under this lens, the energy crisis appears as the outcome not only of oil policy, production, and consumption but also of the material and cultural history of the postwar United States, including Cold War fears over energy resources and population growth, the development and popularization of ecology, and new experiences of the flow of time rooted in automobility. It builds on Huber's insight that neoliberalism flourished within an atomized vision of life enabled by fossil fuels to show how energy crisis narratives from a variety of sectors—environmentalists, economists, populists, artists, and others—mobilized the energy crisis in ways that aided neoliberalism, including positing the market as the only alternative of keeping the government from controlling how Americans used energy and the necessity of speculative markets to retain control over energy pricing. The energy crisis was discursive, which was the only way to bring a crisis of energy into being, since we do not actually *see* stores of oil, let alone energy. To articulate a crisis based on the shortage of a virtually invisible commodity is to rely on surveys, forecasts, and discourses that frame the experience of the present as portents of something deeper. Those competing narratives helped to energize neoliberal visions of renewed abundance through free markets.

This book is organized chronologically and thematically to explore how the energy crisis was assembled, interpreted, acted on, resisted, and eventually "resolved." The story of the energy crisis begins in the first chapter with the practices of mass consumption that made up US petroculture, which emerged after 1945 as an integral part of the Cold War politics of capitalism after years of economic depression and war. Efforts to shore up an anti-communist American "way of life" devoured energy, producing extensive dependencies on fossil fuels and, crucially, the *expectation* of an abundant energy future. Yet anxieties around energy, resource scarcity, and nuclear apocalypse shadowed America's postwar energy binge. The concept of energy evolved into a flow through systems around the same time that the US government developed a coherent notion of a US energy system in need of management, creating concern about its future. Concurrently, the revival of Malthusian environmentalism and the nuclear imagination of disaster unnerved how many Americans felt about the future of their consumers' republic. The chapter concludes by considering how these habits of consumption and thought joined the proliferation of crisis discourses in the 1960s and 1970s, and disparate concerns about energy supplies in that period, to become a national crisis of energy in the early 1970s. When the oil embargo struck in late 1973, many Americans experienced the crisis that had already been publicly discussed for months in the form of gas lines, inflation, and forecasts, anticipating even greater interruptions in the fossil-fueled flow of time and national vitality in the years ahead.

The rest of the book explores the energy crisis as an interpretive struggle at four key sites: expert debates about the meaning of the energy crisis, efforts to establish a conservation ethic in individual citizens, petro-populist representations of automobility in 1970s US cinema, and the introduction of oil futures trading on the New York Mercantile Exchange in the late 1970s and early 1980s. The account in chapter 2 of the debate about the meaning of the energy crisis shows that it was primarily understood in three ways: as a sign of national moral failure, as a marker of transition, and as an injunction to imagine new futures. The political flexibility of "crisis" is clearest in this chapter and the next, which both consider how crisis claims informed neoliberal and environmentalist visions of the future, to name but two. The third chapter does this by reconsidering the deceptively bipartisan support that energy conservation attracted as a solution to the crisis, which generated support for market deregulation to discipline consumer behavior. The conflict over energy conservation included debates about individual freedom, consumer behavior, and the regulatory role of government that played an important part in the nation's rightward shift.

The final two chapters note the rise of a neoliberal political culture as it related to energy. The fourth chapter pivots from energy conservation to the populist backlash against it by examining the neoliberal structure of feeling of 1970s "car films." During the energy crisis, dozens of films about working- and middle-class men who rebel against speed limits, environmentalists, and feckless sheriffs articulated a petro-populist view of US petroculture and individual freedom within it that largely rejected energy conservation as an anathema. Turning from popular culture to economic discourse, which I treat as relevant to each other, the final chapter demonstrates the consolidation of a neoliberal socioeconomic paradigm in the evolution of oil futures trading on the New York Mercantile Exchange. The turn to futures trading to bring control over oil prices back to US markets, while generating profits to fund domestic oil exploration, demonstrates the victory of a market conception of the energy crisis over an environmentalist conception. Arguments and representations favorable to futures markets, moreover, framed them in populist terms as democratic spaces where "the people" decided the price of goods, rather than government fiat. The epilogue traces the echoes and legacies of the energy crisis at key moments, including the financial panic surrounding the Gulf War, when futures trading worked against US interests. The themes of crisis, petroculture, and neoliberalism that shaped the energy crisis continue to haunt our contemporary politics of climate change and capitalism, which are similarly anticipatory crises that mask competing visions of the future.

This book argues that 1970s energy crisis narratives supported the emergence of neoliberalism in the United States, while also attending to the flexibility and durability of crisis as a mode of analysis. It uses the energy crisis to trace the normalization of a politics of crisis and anticipation and the complex role of an ecological critique in US petroculture while broadening the historiography of America's turn to markets in the late twentieth century. It offers an interdisciplinary account of the emergence of neoliberal subjectivities in relation to energy scarcity, and it suggests that crisis is an insufficient foundation for a viable politics of energy transition. It invites readers to ask how we might know life beyond markets and crises.

Chapter One

"Is America Running Out of Gas?"

Assembling the Energy Crisis

As a Nation, we are threatened, but not alert.
—THE PRESIDENT'S MATERIALS POLICY COMMISSION, 1952

On its January 1974 cover, *Time* asked if the energy crisis was "real or phony." The question reflected the broader skepticism of the American public toward the energy crisis. Many people refused to believe that it was "real," that there were actual shortages of oil behind rising prices and depleted gas stations. Rumors swirled about gasoline price gouging, refineries awash in crude, and oil tankers waiting offshore until the price was right. Critics accused OPEC, oil companies, or the government of contriving scarcity by withholding oil or failing to manage its production. Some called for economic or military retaliation against OAPEC for its use of the "oil weapon," and many demanded the breakup of monopolistic oil companies to restore price-cutting competition.[1] By April 1974, Gallup reported that 74 percent of Americans thought the energy crisis to be "phoney."[2]

Populist skepticism and apathy toward the energy crisis ebbed and flowed throughout the 1970s. This frustrated government officials struggling to formulate coherent policies for a polity growing increasingly cynical about their competence to manage the economy. As William E. Simon, a Wall Street investor and key member of the Nixon and Ford administrations, confided to a prominent businessman in 1974, the public had to accept "that the crisis is not only real but has been with us, unacknowledged, for years."[3] Simon frequently appeared in the media to attest to this fact, but public apathy persisted, particularly as the embargo ended.[4] And so President Jimmy Carter and his officials also fretted in 1977 about communicating the reality and urgency of the energy crisis, particularly in order to pass his cornerstone National Energy Plan (NEP).[5] The text of the NEP warned that it was easy to forget, "in the absence of energy traumas," that the end of the oil embargo had not resolved the crisis.[6] Pollster Daniel Yankelovich explained to Carter that the crisis seemed "remote and unreal" because there were "no visible signs of it" and the president was "not talking about it any more." Carter's initial energy addresses, he said, "were like the first tap of a hammer on a nail being driven into a very hard piece of wood."[7] Carter spent the rest of his presidency talking about the energy crisis, publicly addressing it at least seventy times.

Yankelovich's insight that popular belief in the energy crisis required both "visible signs" and "talking about it" hints at its origins, as well as at the nature of crisis as a form of political knowledge making. As Janet Roitman argues, a crisis must be "constituted as an object of knowledge" that, once articulated, can be used politically.[8] But how are crises constituted? In this chapter, I argue that crises are *assemblages* of material and discursive elements that cohere in specific historical circumstances.[9] The history of fossil fuels in the United States is littered with controversies and anxieties over resource supplies, prices, and production levels. But none of those controversies congealed as energy crises because they lacked the material and discursive preconditions that led to crisis in the 1970s. Although the 1970s crisis was often disputed as phony, it mattered in US political culture as a site of interpretation, contestation, and action.

This account of the 1970s energy crisis begins with the creation of a petroculture within the Cold War politics of capitalism. Although certain prewar developments enabled deep fossil fuel dependence, the postwar period is uniquely important to understanding the genesis of the energy crisis. In the wake of the Great Depression and Second World War, the United States realized a level of energy consumption unprecedented in human history that intensified pressure on natural systems. This high-energy society created distance from the deprivations and sacrifices of the 1930s and 1940s, but its motive force came from the political imperative to demonstrate the superiority of capitalism over communism. The Cold War therefore completed oil's transformation from minor illuminative commodity to indispensable fuel for heat, light, motion, and materials. It shaped the materiality of US petroculture, at the center of which lived a decentralized, majority white, and increasingly suburban lower- and middle-class population reliant on automobiles, appliances, asphalt roads, and synthetic materials. The energy abundance that drove this transformation came to be taken for granted, producing the expectation of cheap oil and its associated capacities.

But the nation's postwar energy binge was not without its troubles. The rapid growth of a high-energy society generated anxieties that informed the discernment of an energy crisis in the 1970s. The middle sections of this chapter focus on three of them—energy, scarcity, and apocalypse—to explore the key conceptual and discursive conditions of possibility for the energy crisis. The concept of energy shifted in scientific and political discourse from energy as a universal force channeled through laboring bodies and machines to a manageable flow through systems in the work of postwar systems ecologists and the Paley Commission (1952). Concurrently, the Cold War revived Malthusian environmentalism, which articulated fears of apocalyptic environmental collapse and the ever-present threat of devastating resource scarcity that shifted from overpopulation and food scarcity to energy in the early 1970s.

This chapter's final section explores the assembling of an energy crisis in the years before the oil embargo. Experts drew on the developments and categories of energy, scarcity, and apocalypse to narrate a disparate series of events, from electricity brownouts to peaking oil production, as signs of a deeper crisis. A sense of futurity was important: many early claims of an energy crisis referred either to unfulfilled expectations about the experience of time or to the anticipation of worsening scarcity in the future. The claim of crisis intensified when many Americans experienced energy scarcity in the first months of the embargo, in the form of gas lines and inflated prices, but it was the larger postwar context that shaped understandings of those experiences as a sign of things to come. The 1970s energy crisis was thus a historically specific assemblage unique to the late twentieth-century United States.

Crisis Foundations: Modernity, Energy, Acceleration

The postwar energy crisis assemblage built on several long-term foundations. The first was a new configuration of time within modernity.[10] Reinhart Koselleck dates the birth of modernity to the sixteenth century, when time came to be understood as a transformative historical agent rather than the space in which history happens. The passage of time, that is, became linear and implied change rather than unfolding through relatively constant cycles. This new sense of time separated what Koselleck calls the "space of experience" from the "horizon of expectations": increasingly, the experience of the past could no longer reliably guide ideas about the future.[11] Modernist thinkers and political movements, such as Karl Marx and communist movements, intensified this divide by supposing that "progress" should and would ultimately mark the difference between past and present as the pace of historical change accelerated.[12] In modern societies, expectations consistently exceed past experience with the proliferation of the new resulting in a suspended state of anticipation for the future and a need to bridge the widening gap between experience and the demands of the future-in-the-present. The expected difference between past and future is the wellspring of crisis.[13] In the twentieth century, this expectant orientation evolved into a series of preemptive practices as the need to manage the gap between experience and expectation became more urgent.[14] Without this sense of time and the injunction to preempt, the energy crisis would have been unthinkable.

If the relatively abstract sense of modernity was a *longue durée* foundation for the energy crisis, several more concrete historical developments in the late nineteenth and early twentieth centuries set the stage for the emergence of a petroculture in the postwar United States. First, fossil fuel consumption grew rapidly and utterly transformed the country in the space of half a century. At the time of the American Revolution, American energy use consisted largely of fuelwood and muscle power. By the mid-nineteenth century, coal fueled westward expansion via railroads and America's growing industrial capacity. Energy demand nearly quadrupled between 1880 and 1918, mostly in coal, which remained the dominant fuel source at the end of the First World War.[15] The discovery of petroleum in Pennsylvania in 1859 initially only disrupted the use of whale oil for illumination, but the invention of the combustion engine helped oil to surpass coal in 1949 as the dominant fuel in America.[16] Crucial to this process, as Christopher Jones has shown, was the mass construction of energy-delivery infrastructures, which were largely in place by 1930.[17] Fossil fuel consumption benefited as well from mass roadbuilding in the interwar period that encouraged the early adoption of private automobiles and mass electrification, which dramatically altered lives illuminated.[18] Given its long ties to coal, electricity was yet another expression of an emerging fossil economy. As late as 2021, fossil fuels still accounted for nearly 61 percent of all electricity generated in the United States.[19]

This foundation of rising energy abundance rooted in fossil fuels profoundly altered US society. Crucially, it elevated consumption to a more central role in the lives and aspirations of many people. The Great Depression nearly derailed that project, but Roosevelt's New Deal saved it. As Matthew T. Huber has shown, the New Deal's reorientation of US political culture set the stage for even further growth in energy consumption in the postwar period in its vision of the "American Way of Life."[20] New Deal programs designed to rescue capitalism from its greatest economic crisis sought to create a stable market for affordable oil. The oil would then fuel the mass consumption that the government pledged to a nation destined to be homeowners earning high wages. The New Deal's promise of abundant opportunities for consumption "entrenched" the oil-soaked "geographies of social development that we still live with today."[21]

Although these developments formed some of the deep roots of the 1970s energy crisis, the postwar period is uniquely significant for coming to grips with why claims of an energy crisis emerged when they did. After 1945, the world witnessed an unprecedented growth in energy consumption and broader human impacts on the planet's resources. As J. R. McNeill and Peter Engelke note, that growth was so great that the post-1945 era has earned the moniker "the Great Acceleration." In short order, the global motor vehicle population grew by more than 2,000 percent while its human population nearly tripled and moved en masse to cities and suburbs made of asphalt, concrete, glass, and plastic.[22] These and other developments changed how many people lived and how they perceived the world and their place in it. In many ways, the United States led the Great Acceleration on its way to the 1970s energy crisis. It is to that process of material transformation that we now turn.

Producing Scarcity: The Postwar Regime of Mass Consumption

US petroculture and its unprecedented regime of mass energy consumption developed most intensely after the Second World War. The process drew strength from rural depopulation, the decline of central cities, and the proliferation of low-density suburbs that became the "landscape of mass consumption" on which the energy crisis would eventually be constructed.[23] In the aftermath of the Great Depression and the sacrifices of the Second World War, millions of Americans bought single-family homes in burgeoning suburban communities, swelling the suburban population by 43 percent between 1947 and 1953.[24] Similar growth continued throughout the postwar years, aided by various government subsidies.[25] Housing starts took off and held at one million homes per year throughout the fifties and sixties before plunging in the mid-1970s.[26] Built on inexpensive, peripheral land that could accommodate low-cost construction methods, suburbs made private homeownership widely available for the first time in low-density communities marked by economic, racial, and aesthetic homogeneity. In contrast to Europe, suburbs in America sprawled outward, increasing the distance between homes, schools, churches, and shopping as they consumed meadows, wetlands, and farmland. Detached homes accounted for nearly all new single-family homes built in the first decade after the war, exacerbating the problem of sprawl.[27]

Suburban life created a regime of mass consumption that remade social and political life.[28] Encouraged by advertisers and the federal government, new homeowners became mass consumers, filling their homes and garages with electric appliances, gas-powered lawnmowers, and many other consumer items. In the first five years after the war, spending on appliances and other home furnishings grew 240 percent as the home became the center of consumption.[29] Americans bought 20 million refrigerators, 5.5 million stoves, and 11.6 million televisions.[30] When the housing market eventually slowed, homebuilders introduced energy-guzzling features like air-conditioning to attract new buyers.[31] As prominent journalist and urbanist William H. Whyte Jr. observed in 1956, "As a normal part of life, thrift is now un-American."[32]

The Cold War politics of capitalism catalyzed postwar suburbanization and the regime of mass consumption that accompanied it.[33] These were not natural or inevitable developments. The federal government viewed private homeownership as a vaccine against the threat of communism in a chaotic postwar world, so it actively encouraged landownership and participation in a culture of mass consumption, as symbolized by the single-family suburban home armed with an array of appliances and other products. The home symbolized the efficacy of capitalism and, more importantly, gave many middle- and working-class Americans a financial stake in its reproduction.[34] As real estate tycoon William J. Levitt noted in 1948, "No man who owns his own house and lot can be a Communist. He has too much to do."[35] Vice President Richard Nixon's famous "kitchen debate" with Soviet Premier Nikita Khrushchev at the 1959 American National Exhibition in Moscow further cemented this ideological framework of capitalist superiority through the freedom to consume.[36]

Building a suburban defense against Soviet collectivism required access to private automobiles for work, shopping, and entertainment, though the car quickly became much more than a "necessity."[37] In the first five years after the war, Americans bought 21.4 million cars, and throughout the postwar period the rate of increase for automobiles outpaced population growth during America's much ballyhooed baby boom by a wide margin.[38] Annual new car sales averaged nearly 6 million throughout the 1950s, and by 1960, 77 percent of American families owned an automobile, compared to 51 percent in 1941. Such growth swelled the volume of traffic from 200 billion vehicle miles in 1945 to 588 billion in 1960.[39] It also reshaped the economy: by the time US car production peaked in 1965 at 11.1 million vehicles per year, the automotive industry accounted for one out of six American jobs.[40] For its part, the federal government undertook "the largest public works project in human history" by building the Interstate Highway System between 1956 and 1975: a network of more than 42,500 miles of highways costing more than $100 billion to improve mobility and safety while contributing to economic growth and Cold War national security.[41] The built environment also evolved to support America's "four-wheeled love affair" with roadside motels, gas stations, drive-in theaters, and massive parking lots.[42] The energy crisis would be unthinkable without this network of suburbs linked together by an expanding system of highways, roads, and cars.

The rapid growth of US suburbs and automobility constituted a regime of energy consumption unprecedented in human history. National energy consumption skyrocketed as more and more people used electric lighting and appliances, heated private homes, mowed lawns, purchased synthetic consumer goods, and drove cars to work and vacation. Between 1945 and 1973, total US energy consumption measured in BTUs (British thermal units) grew by nearly 130 percent. Oil and natural gas consumption drove growth, increasing 230 percent and 467 percent, respectively. Coal declined by 18 percent, but that figure is exaggerated because its use had spiked for war production.[43] By 1973 fossil fuels made up most US energy consumption, driving economic growth and normalizing high-energy living. Fossil fuels also accounted for nearly 81 percent of electricity generation.[44] Aside from brief retreats immediately after the oil shocks in 1973 and 1979 and the Gulf War in 1991, US energy consumption continued its precipitous climb for the remainder of the century, even as manufacturing struggled and service industry jobs multiplied.

The development of what David Nye calls a "high-energy society" extended far beyond the borders of suburbia.[45] By the middle of the century, agriculture depended on fossil fuels to power tractors and to fertilize crops. Transport trucks replaced trains to haul goods from state to state on the nation's growing network of highways.[46] In major cities, gasoline-powered buses replaced electrified streetcars, and in the Northeast, oil replaced wood to heat homes in cities, towns, and suburbs. These changes rendered individuals, families, and communities dependent on far-flung networks of energy extraction and transportation. At the same time, the petrochemical industry introduced dozens of new materials that tightened the cultural and economic grip of fossil fuels.[47] This sea change in materials famously inspired the career advice offered to the aimless young protagonist of *The Graduate* (1967): "Plastics . . . there's a great future in plastics."[48]

This emerging, car-centric petroculture altered the experience of time and space for millions of Americans from all classes and races, though the effect was uneven.[49] Driving miles rose by over 1,900 percent between 1921 and 1970.[50] Travel became more individualized as passenger volume on trains and subways declined. The spread of rail travel had compressed time and space in a hierarchical and linear fashion, but cars and the growing network of roadways accelerated and individualized that sense of compression.[51] American drivers, that is, began to drive more often, to different places, and on their own schedules. They also moved faster. Between 1945 and 1970, the average highway speed of passenger cars increased from 44 mph to 61 mph, a 39 percent jump.[52] Advertisements embodied the growing importance of speed by substituting excitement for older commercial associations of cars with luxury, class status, and practicality. By the time the French philosopher Jean Baudrillard toured the United States in the mid-1970s, he remarked on the nation's absolute commitment to speed as "the triumph of instantaneity over time as depth."[53]

The car-centric design of suburban landscapes also reshaped the rhythms of daily life for many. Low-density suburbs redefined notions of locality and, by privileging automobiles over pedestrians, ensured that cars were the only safe and convenient mode of transportation for daily tasks.[54] Daily patterns of work and home life changed, as well as weekend leisure time, normalizing the privatized speed of experience afforded by cars. Mass automobility altered what futurist Alvin Toffler called the "durational expectations" of everyday life, that is, expectations about how long different tasks and trips should take to complete. People increasingly experienced cities, neighborhoods, and landscapes at speed, behind the wheel of a car, providing a thrilling sense of self-possession and freedom. Anxiety about this faster, more impersonal "pace of life" informed Toffler's 1970 bestseller, *Future Shock*, which warned about its disorienting effects.[55] Rapid change, he thought, fostered the transience apparent in Americans' growing car ownership and new patterns of movement. He noted that the average American of 1970 traveled nearly ten times more miles per year than in 1914, as new roads accommodated new cars, begetting greater distances between home and work and greater speeds to make up the distance.[56] Driven by New Deal promises and Cold War politics, suburbs thus begat mass consumption and mass automobility, which consumed enormous amounts of energy and created rhythms of life that depended on cars and the cheap oil that fueled them.

The popularity, ideology, and materiality of American suburbs and of the larger high-energy economy of which they were a part fostered reliance on and the expectation of cheap energy, even as actual interactions with energy became more abstract and impersonal. Few people ever saw the gasoline that powered their cars, but they depended on its availability and the capacities that it afforded. The same was true for the other forms of energy supporting postwar life. Rather than firewood or coal, natural gas and oil heated increasing numbers of homes and fossil-fueled electric air-conditioning cooled them, flattening seasonality into a smaller range of temperatures and bodily expectations about appropriate degrees of warmth and cold. Tractors and fertilizers wove fossil fuels so tightly and invisibly into the fabric of farming that ecologist H. T. Odum could claim in 1971: "Industrial man no longer eats potatoes made from solar energy, now he eats potatoes partly made of oil."[57] As it devoured huge quantities of oil, gas, and hydropower, America's postwar boom changed expectations about standards of living.[58]

Geopolitical and policy dynamics exacerbated America's growing dependence on fossil fuels. First, postwar energy policy taxed domestic oil resources to their limits. Although the United States was self-sufficient in coal and imported negligible amounts of natural gas, neither of which faced competition from cheaper foreign sources, oil abundance and falling prices worried the US government and independent oil producers.[59] The latter wanted protection from cheap Middle Eastern and Venezuelan oil. They managed to convince the federal government that reliance on foreign oil threatened the integrity of the national oil economy and national security, leading the Eisenhower administration to establish oil import controls in 1959. The Mandatory Oil Import Quota Program capped imported oil at about 12 percent of the American market, reserving the rest for American producers.[60] Quotas sheltered the American market from foreign competition for fourteen years, accelerating pressure on US oil reserves by capping the amount of foreign oil that Americans could consume. This system, however, was viable only if US oil resources could satisfy US oil demand. When oil domestic production began to decline and spare capacity evaporated in the early 1970s, US oil imports began to climb.[61] President Nixon lifted import restrictions in 1973 to keep the US economy awash in oil, but his environmental achievements and penchant for price controls, which he calculated would help him win a second term, continued to frustrate efforts to ramp up domestic production.[62]

While America's domestic oil flows dwindled, its companies began to lose control over foreign oil flows. By the middle of the twentieth century, anti-colonial nationalist movements had gathered enough strength to threaten the system of concessions that had been enriching Western oil companies at the expense of their host countries.[63] The people and leaders of oil-producing countries such as Iran, Iraq, and Saudi Arabia resented what they took to be an outdated and unfair relationship with Western oil companies and wanted more control over the price and profits of their oil. Mexico was the first to nationalize its oil industry in the late 1930s, and beginning in 1948, other oil-producing nations began to demand new concessions agreements for a 50/50 share of profits.[64] When British Petroleum refused to grant Iran a 50/50 agreement, the country attempted nationalization in 1951, but it was prevented from doing so by an embargo by the major international oil companies (the Seven Sisters) and a coup against its nationalist prime minister, Mohammad Mossadegh.[65] Nevertheless, the 50/50 agreements and the rising tide of nationalism prefigured a changing international oil industry.

Later in the decade, spurred on by repeated and unwanted price cuts imposed by Western oil companies trying to stabilize global oil flows, Iran, Iraq, Kuwait, Venezuela, and Saudi Arabia formed OPEC—the Organization of the Petroleum Exporting Countries—in September 1960.[66] The organization's original goal was to ensure that producing countries received a fair share of oil rent from the Seven Sisters.[67] OPEC struggled to agree on much else during its first decade, and most Western governments and companies did not fear it, particularly after a failed attempt at an oil embargo in 1967 during the Six-Day War. But as global oil demand continued to rise while US production stalled, the conditions were ripe for OPEC to capitalize on a global oil market quickly turning in its favor. Led by Libya in 1971, OPEC nations began to renegotiate concessions again, setting a new minimum of 55%–45% in their favor.[68] As it continued to retain more influence over oil prices, the newly emboldened cartel was poised to leverage its oil for other purposes.

Anxieties of Abundance

America's regime of mass consumption was not without its anxieties.[69] While oil companies and the federal government tried to manage surpluses, the combination of lifestyles based on mass consumption, spectacular global population growth, and Cold War security matters stirred anxieties about catastrophic resource scarcity. Such concerns nurtured the nascent environmental movement and crucially provided an interpretive frame for early 1970s energy shortages. Claims of an energy crisis were not merely the recognition of geological or geopolitical "facts" about oil reserves and energy policy. They drew on key postwar discourses and imaginaries of energy, scarcity, and apocalypse.

Constructing an Energy System

America's energy crisis assemblage presupposed a coherent energy system as the site of crisis. Timothy Mitchell first made this point in *Carbon Democracy*, where he argues that the process of thinking about "energy" as a coherent object required the conceptualization of an energy system. He attributes this development to the emergence of "energy companies" in the early 1970s.[70] By controlling multiple energy resources, oil companies could demand higher land leases and raise prices while using inflated forecasts of future demand to foment concern about future energy supplies.[71] The White House further facilitated this process by creating federal agencies that consolidated oversight of different fuels into a single energy agency. Oil company rhetoric about the environment, economic discourses of supply and demand applied to energy shortages, and the sudden emergence of an oil crisis linked to Middle East politics also helped to create "energy" as a coherent field. What Mitchell's account leaves out are the postwar developments that set the conditions for this rather swift systemization and consolidation of "energy" in the early 1970s—specifically, an ecological notion of energy and the postwar energy resource surveys undertaken for national security that defined the energy system.

The systemization of energy first required a new understanding of energy as an input and animating property of systems. In the nineteenth century, the new science of thermodynamics posited the cosmos as a reservoir of productive power awaiting conversion into productive work through bodies and machines.[72] It supported an emerging science of work that sought to manage human bodies and machines for maximal productive efficiency. The twentieth-century development of the ecosystem concept in American ecology shifted notions of energy from the laboring body and machines to systems. Energy as flow emerged to replace energy as productive power, a notion well-suited to an increasingly technocratic, networked society.[73]

The concept of an "ecosystem" facilitated the reconceptualization of energy by framing it as the animating feature of systems and their development over time. Three American ecologists, all of whom studied under the Yale limnologist G. Evelyn Hutchinson, contributed to this development by theorizing and popularizing ecological systems in terms of energy flows: Raymond Lindeman, Eugene

Odum, and his brother Howard Thomas Odum.[74] Lindeman's contribution, published in 1942, was to link the ecosystem as the basic unit of ecology to energy as its currency. In this, Lindeman drew on the work of the British plant ecologist Arthur Tansley, who had introduced the ecosystem concept in 1935 but did not see energy as a central feature.[75] As he studied Cedar Bog Lake, a small, senescent lake in Minnesota, Lindeman differentiated energy from matter and posited that solar energy flow drove the dynamics of the system.[76] Indeed, his theory suggested that energy flow was the fundamental process in the developmental life of ecosystems, driving succession over time.[77] As Gregg Mitman notes, Lindeman's conceptualization of ecosystems as "a system of energy flow from one trophic level to the next" facilitated a paradigm shift from the cooperative, organicist orientation of community ecologists toward focus on competition between species.[78]

The ecologists and brothers Eugene Odum and H. T. Odum further led the development of systems ecology in postwar America, as well as popularizing the ecosystem concept and its sense of energy as a flow through those systems. In the mid-1950s, they completed a groundbreaking study on ecosystem functioning using a US nuclear weapon testing site, Eniwetok Atoll. That study, and several others, was funded by the Atomic Energy Commission, a key department of America's Cold War effort interested in understanding how nuclear radiation moved through natural systems.[79] Building on methods developed by H. T. Odum, the Odum brothers measured what they called the "metabolism" of the atoll's ecosystem powered by solar energy flows. The brothers also wrote a phenomenally successful textbook, *Fundamentals of Ecology*, that helped the ecosystem gain currency far beyond academia to become part of the environmentalist lexicon in the 1960s.[80] Eugene Odum went on to become an influential advocate of the ecosystem concept. He argued that the problem of human population growth drove the study of energy flow through ecosystems, suggesting overlap between neo-Malthusian concerns and ecological science in the formation of an energy crisis assemblage.[81]

The Odums' notion of ecosystems as interacting communities of living and nonliving organisms united by the circulation of energy facilitated the shift to energy flow as a crucial input into any system.[82] H. T. Odum, who developed a more mechanistic and energetic outlook than his brother, drew on the 1920s work of the mathematician and chemist Alfred J. Lotka to argue that all systems are subject to the laws of thermodynamics.[83] Energy flow, he thought, is the common denominator to all systems and the key to understanding how they work and how best to manage them.[84] By conceptualizing energy in this way, the Odums built the tools by which the energy that animated human social and economic systems could be conceived of in systems terms as a coherent flow subject to expert management. As the *New York Times* remarked at the nadir of the oil embargo, energy is "not just a commodity to be bought and sold like any other, it is a central component in the functioning of the society."[85] Energy itself had become a site of social and political intervention.

While the Odums developed theories about ecosystems and energy flow, energy became further legible as a property of systems in the work of humanist scholars who theorized the impact of differential energy flows on human systems. Largely ignored in their day, the work of anthropologist Leslie White and sociologist Fred Cottrell demonstrates how notions of energy as input and flow through systems moved across disciplines in the postwar years. Near the beginning of the Atomic Age, White brought an ecological and systems view of energy into the study of human societies. He posited energy flow as the motor of cultural development much in the same way that it drove ecological succession for Lindeman. "Everything in the universe," White wrote in 1943, "may be described in terms of energy," including atoms, organisms, and societies.[86] The evolution of a given culture depended entirely on its ability to harness increasing amounts of energy. In a later work, White specified cultures in systems terms, calling them "dynamic systems [that] require energy for their activation" and that should be thought of as "a mechanism for harnessing energy."[87] This concept helped to shift notions of energy from the embodied to the external input into systems that make them run. Crucially, his notion of energy was abstract and did not differentiate between different sources. The important thing was to harness more of it. Looking to the future, White concluded: "If we can continue to harness as much energy per capita per year in the future as we are doing now, there is little doubt that our social system will give way to . . . a new era of civilization."[88] White tried to account for differences between energy sources by positing that social systems tend to support magnitudes of energy commensurate with the sources that they are developed to harness, but the key to progress lay in amounts of energy rather than forms.[89]

Dominic Boyer has argued that White's work hints at the idea that modern capitalism is a "fuel society to its core," dependent on extensive consumption.[90] An insight of that nature did appear during White's lifetime, however, in Fred Cottrell's *Energy and Society* (1955), which argued that energy shapes the organization of human cultures, as well as their political, economic, social, and psychic features.[91] At low levels of energy development, human behavior accords with a limited range of possibilities that evolve in tandem with new sources of energy.

Capitalism, Cottrell argued, required high energy yields and tended to be organized to reproduce them by centralizing power.[92] Both White and Cottrell, then, conceived of energy as a fundamental, external input to human systems that set limits and generated new possibilities. Their achievements suggest that the Cold War and nuclear power fostered new forms of systems thinking in ecology and in the social sciences about the relations between energy and systems. Indeed, Cottrell and White worried about the future given Cold War realities—Cottrell about nuclear power and White about diminishing fossil fuel resources. Similar concerns about the status and effects of energy systems informed the resource surveys of the postwar period, to which we now turn as the final step of the systemization of energy into a coherent system.[93]

Beginning in the late 1930s, the US government conducted energy resource surveys that began the process of calculating a US energy system and its prospects. These surveys continued in the postwar period and were driven by emergent anxieties about the sustainability of a mass consumer society and Malthusian population growth, as well as the ideological and military imperatives of the Cold War.[94] The most prominent was the Paley Commission, established by the federal government in 1952 to inventory and forecast US and free world natural resources up to 1975. By figuring out how many resources the nation had, how many it would need, and how best to provide sufficient energy and other resources in the future, it was hoped that the system could be managed to support the long-term national security and economic vitality of the United States and its allies.[95] The Paley Commission also affirmed its commitment to the main tenants of Cold War capitalism: economic growth, private enterprise, and the shared economic and security interests of the US sphere of influence. Resource scarcities threatened each of them.[96] The report revealed particular concern for the possibility of future wars and the emerging "pattern of world oil that is taking shape." It feared that dependence on imported oil from the Middle East posed a grave "threat to the wartime security of the free world."[97] In accounting for its energy resources, the US government hoped to plan for its continued geopolitical and economic dominance in the twentieth century.[98]

Mitchell claims that "in the 1950s and 1960s, the concern with oil and other fuels was usually part of a general issue of 'natural resources,' with postwar fears put to rest by . . . the Paley Commission."[99] Yet in its concern about the consequences of future wars on US resources and its uncertainty about the possibility of war, the Paley Commission embodied deeper anxieties than Mitchell allows. The report laid out in clear terms the resource vulnerabilities of the United States, as well as their fundamental strategic and economic importance. It predicted a range of scarcities in minerals and energy by the 1970s and urged "investment in security," including more free trade with allies.[100] Most importantly, although it explored energy futures under the heading of "resources," the Paley Commission, like other studies from this period, demonstrated a keen understanding of the interconnectedness of different energy resources and the importance of managing them as a unit with unique value to the wealth and security of the nation.

Studies like the Paley Commission generated a sense of an energy system by developing an abstract and holistic notion of "energy resources" as a distinct category of resources in an interconnected system of forms. Energy surveys and forecasts constructed the US energy system using an ecological sense of "energy," conceptualized and calculated by the Paley Commission as "total energy."[101] So while it is true that energy was studied under the rubric of "resources" in the postwar period, energy resources were calculated and mapped as a totality of energy available to the state and national economy, bringing those diverse resources into a specified field of energy. This was an ongoing process. The Paley Commission used data from a 1939 energy resources report that Franklin Roosevelt introduced by noting:

> In the past the Federal Government and the States have undertaken various measures to conserve our heritage in these resources. In general, however, each of these efforts has been directed toward the problem in a single field: Toward the protection of public interest in the power of flowing water . . . toward the relief of economic and human distress in the mining of coal; or toward the correction of demoralizing and wasteful practices and conditions in the industries producing oil and natural gas. It is time now to take a larger view: To recognize—more fully than has been possible or perhaps needful in the past—that each of our great natural resources of energy affects the others.[102]

As early as early 1939 a coherent system of energy was also coming into being.[103]

The process of surveying and forecasting in the Paley Commission built on earlier studies by stressing "the strong interrelationships among energy sources" and the necessity of cheap and abundant energy for economic growth and national security. The report operated with a sense of a unified energy system when it contended that "all sources of energy must be considered not as separate entities but as the related parts of an essential whole."[104] Indeed, it said that the future of US energy security relied on understanding "energy resources as a whole" in order to create a dynamic and sustainable energy mix.[105] Under the rubric of "total energy" the report discussed coal, natural gas, oil, and hydro as the constituents of one system. Exhaustion was central to this conception of an "energy mix," which it developed through the notion that fossil fuels might eventually fail to meet "civilization's need for energy," forcing reliance on nuclear or solar power.[106] Here, again, was a system composed of different forms of energy under an abstract sense of the "energy" necessary for the reproduction of a capitalist political and economic order.

Ultimately, the Paley Commission asserted its confidence in capitalism to procure enough energy for the medium term, but its conceptualization of energy resources as the "total energy" set the stage for later resource surveys and works of synthesis that further cemented the notion of a coherent energy system essential to US capitalism.[107] The Shell petroleum geologist M. King Hubbert, famous for predicting that US oil production would peak in 1970, received funding from the National Academy of Sciences–National Research Council in 1956 to survey the status and prospects for several energy forms, including oil, coal, natural gas, shale/tar sands, nuclear, solar, tidal, and geothermal energy. Even more than the Paley Commission, Hubbert framed his survey with the language of ecology and systems theory, noting that the earth is "a material system" characterized by a "continuous flux of energy."[108] His "Energy Flow Sheet for the Earth" exemplified his coherent, systems view of the sources of energy available to Americans by visualizing the energy system as an interconnected series of flows.[109] Like all neo-Malthusians, however, he saw the downside of intensive energy exploitation in unchecked human population growth.[110]

The discursive and calculative process of producing a US energy system continued in the 1960s with the publication of still more surveys. In its landmark 1960 study *Energy in the American Economy*, the think tank Resources for the Future (RFF) compiled data about "total energy consumption" in the past, present, and future while surveying the history of the US energy system through the linked histories of oil, coal, fuelwood, and natural gas.[111] A few years later, the Office of Science and Technology hired RFF to study the US energy situation to support the coordination of energy policies for the entire government. Like the Paley Commission, RFF concluded that the interchangeability of energy resources meant policy had to consider energy "in the aggregate."[112] This process was aided by a 1971 RFF study that made different forms of energy "statistically comparable" by converting each into a common unit based on "calorific values."[113] Coal, oil, natural gas, solar, and other forms of energy could be aggregated and calculated as the total energy production of a global energy system. The study used these calculations to connect higher energy consumption to higher economic growth and standards of living, irrespective of the forms of energy production in play and their differential social costs. It also produced a new understanding of the global economy and its constituent nations as a global energy system composed of national systems that could be compared, managed, and forecasted. Thus, by 1968, there existed notions of energy as flow through systems and of a coherent system of energy forms, as well as the idea that the federal government needed overarching energy policies to coordinate resource development and conservation. Oil companies played a key role in crisis discourse, but Cold War concerns about the vitality of the capitalist system established these essential frameworks for the energy crisis.

The Neo-Malthusianism Framework of Scarcity

In the late eighteenth century, the clergyman–turned–political economist and demographer Thomas Robert Malthus posited a wicked conundrum: human populations naturally grow geometrically, but food supplies can only increase arithmetically. Without prudent behavior, Malthus thought, these differential rates of increase inevitably create population pressures on the environment, leading to starvation, disease, and death.[114] Malthus called this concept the "population principle," and it went on to inform the vanguard of intellectual life in nineteenth-century Europe, including Darwinian evolution. Early twentieth-century eugenicists also drew on Malthusian morality in debates about reproduction and birth control, and as explanations for persistent poverty in industrializing Europe.[115] Malthus initially had less impact in the United States, where progressives and conservationists emphasized technology and efficiency rather than fallen humanity and individual morality. In the aftermath of the Great Depression and Second World War, however, Malthusian anxieties crept into US public discourse. Amid its material abundance, mid-century America, in the words of the historian Thomas Robertson, had a "Malthusian moment."[116]

Fears of resource exhaustion date back to the nineteenth-century British economist William S. Jevons, who questioned the sustainability of British coal dependence in 1865.[117] Similar concerns periodically reappeared in Western Europe and the United States in the late nineteenth and early twentieth centuries, animating conservationist movements. After the Second World War, a Malthusian and catastrophist view of resource scarcity returned and deeply influenced US environmentalism. Proponents argued that population growth threatened to produce resource scarcities likely to foment global political unrest if left unchecked. The evident resonance of these views suggests the presence of an abiding anxiety shadowing the postwar frenzy of consumption and material abundance, an anxiety that energy shortages seemed to fulfill when they finally appeared in the early 1970s.

Neo-Malthusianism was always a minority view, but it achieved a degree of public prominence in the years immediately after the war.[118] As the economist Kenneth Boulding noted in a 1959 edition of Malthus's work, "the Malthusian specter is not easily laid."[119] Prominent postwar America conservationists William Vogt and Henry Fairfield Osborn Jr., for instance, published popular jeremiads about the destruction of land and mineral resources groaning under the weight of growing human populations.[120] Exploiting apocalyptic fears sparked by the atomic bomb, both men indulged in global environmental catastrophism that prophesied extensive scarcity.[121]

Vogt's brand of neo-Malthusianism in his popular book *Road to Survival* (1948) tied energy to the problems of population growth and resource scarcity. He wrote:

> All the energy with which man refuels himself is derived from the sun. From the instant the struggling sperm meets the waiting ovum until the last drop of blood dribbles out from the dying heart, man is burning fuel from this one source. His power to move, to grow, to think, to digest, and to beget his kind has the same origin. The hydroelectric power that he uses to reduce the drain on bodily energy—and to make available the energy locked in the atom—is drawn from falling water that was lifted to the clouds by the heat of the sun. The wind that raises water to his farm tanks blows because the heat of the sun causes inequalities in atmospheric pressure on the surface of the earth. Coal and petroleum hold only solar energy captured by plants millions of years ago. Even the great forces in the nucleus of the atom were hurled off from the sun at the world's birth.[122]

Here, energy is a property of systems, manifest in many forms, understood through a rubric of scarcity. Vogt went on to consider the carrying capacities of the land as a limit to human activity because human bodies rely on plants to convert solar energy that, as per Malthus, cannot keep pace with animal populations. His metaphor of the bodily need for energy suggested that civilizations must procure power, but within limits.[123] The book's final chapter, "The History of the Future" appealed for tough political measures based on rising population and declining resources. The continuance of those trends, Vogt warned, could breed global catastrophe, including yet another war, which weighed heavily on the minds of intellectuals in the late 1940s.[124] In the wake of a world war, at the beginning of the Cold War, and amid a burgeoning postwar consumer culture, Vogt, Osborn, and other conservationists warned that large populations with increasing resource demands could breed devastating scarcities that only the prudent, scientific management of population and resources could avoid.[125] The idea continued to inspire key US environmentalists as energy consumption continued its stratospheric rise.

This neo-Malthusian scarcity framework shaped the environmental movement in the 1960s and early 1970s, right up to the declaration of an energy crisis. Although many environmentalists concerned themselves with the negative effects of abundance—too many chemicals, too much air pollution, too many nukes, and too much sprawl—the politics of abundance fed the politics of scarcity.[126] By the end of the sixties, influential environmentalists often linked pollution and resource exhaustion to human population growth.[127] The biologist Paul Ehrlich was the most popular prophet of overpopulation. His polemic *The Population Bomb* (1968) sold more than two million copies and launched his career as a celebrity scientist.[128] In that book, Ehrlich departed from Vogt and Osborn and environmentalists like Rachel Carson by emphasizing the threat that overpopulation posed to the human future.[129] To save the planet and ourselves, he argued for "conscious regulation of human numbers" to achieve zero population growth.[130] In a similarly controversial and influential essay, "The Tragedy of the Commons," ecologist Garrett Hardin warned that human population threatened to exhaust the resources of "a finite world."[131] For these 1960s neo-Malthusians, scarcity was a fact of nature to which humanity must adapt or perish. The audience that they commanded ensured that the Malthusian equation of population and resources had been lodged firmly in public awareness as lights flickered and the oil spigot sputtered in the early seventies.[132]

Neo-Malthusian environmentalism further emphasized resource scarcity in the early 1970s, particularly with the publication of the Club of Rome's controversial *The Limits to Growth* report (1972), which sold ten million copies in thirty languages.[133] Using computer models, the report warned that current patterns of growth—population, consumption, agricultural production—threatened to overshoot the finite carrying capacity of the earth's biological systems, spawning ecological and social collapse. Although *The Limits to Growth* received a hostile reception from many quarters and its authors subsequently moderated their conclusions, it applied the neo-Malthusian framework of scarcity to all resources and popularized the limits to growth concept that informed the energy crisis assemblage.[134] Even Richard Nixon invoked the report's language in his April 1973 energy address.[135] Nixon, though, quickly pivoted from the threat of scarcity to a program of domestic energy production at odds with the aims of environmentalists, suggesting the degree to which the energy crisis united different constituencies with very different interests.

The Specter of Apocalypse

Postwar environmentalists often used catastrophist and apocalyptic imagery to describe humanity's troubled relationship with nature. Perhaps the most important environmental book in American history, Rachel Carson's *Silent Spring* (1962), began with an apocalyptic fable in which "the shadow of death" swept over a small town. Birdsong fell silent, offspring fell ill, and vegetation wilted. A once-verdant countryside now seemed "deserted by all living things."[136] Ehrlich's *Population Bomb*, the title of which adapted the threat of nuclear apocalypse to the threat of overpopulation, also began with a prediction: "In the 1970s and 1980s hundreds of millions of people will starve to death." Ehrlich described efforts to increase food supplies as a mere "stay of execution" because the coming Malthusian crisis was "inevitable."[137] These and other works claimed that the environment teetered on the edge of a disaster that posed an existential threat to all Americans, and to the world at large. The abundance all around could collapse and disappear in an instant.

These forms of what historian Jacob Hamblin calls "catastrophic environmentalism" drew on the politics and imaginaries of the Cold War state.[138] More specifically, they partook of an imagination of disaster made possible by the Cold War and the efforts of the US national security state to exploit the existential threat of nuclear weapons to justify the construction of a military-industrial complex.[139] Many Americans sensed at the moment of their deployment the apocalyptic possibility inherent in nuclear weapons. The atomic bomb spawned anxieties evident in popular culture about the possibility of instantaneous mass death and inspired an antinuclear movement often led by scientists.[140] But the state, as Joseph Masco argues, actively encouraged such morose imagining by routinely using images of nuclear destruction to "militarize U.S. citizens through contemplating the end of the nation-state."[141] The Eisenhower administration launched a mass media campaign to teach public fear of nuclear destruction as a form of psychological preparedness, thereby normalizing the very notion of nuclear apocalypse as a way to inoculate citizens against panic should the unthinkable occur. RAND analyst Herman Kahn's infamous book *On Thermonuclear War* (1960) laid out the possibilities in bone-chilling detail.[142]

Although the apocalypse has ancient roots in the West, particularly through Christianity, the deployment of nuclear weapons and the circulation of images and imaginaries of total and instantaneous nuclear destruction offered a new and visceral way of thinking apocalyptically. Through a barrage of films, photographs, news reports, and exercises, it became possible to think and to feel apocalypse intimately. That such thinking coexisted with Cold War concerns about energy resource scarcity and neo-Malthusian environmentalism is no coincidence. They fed off each other. They created a cultural climate that perceived energy resources as a fragile system necessary to fend of the threat of communism, in which apocalyptic visions of resource scarcity and total systemic collapse spread through bestselling books, television, and political discourse. These were the anxieties that stalked America's postwar energy consumption binge, whispering—or yelling—that it could all be over tomorrow.

The energy crisis assemblage consisted of many discourses, practices, concepts, and relations. In the first half of this chapter, I have focused on some of the material, cultural, and conceptual antecedents of the postwar world that made an energy crisis thinkable, and that informed early declarations of the crisis. Other developments contributed to the assemblage, including US foreign policy toward Israel and the Middle East, environmental legislation, and the emerging sense of national decline linked to the Vietnam War and other national political failures. But these factors played out within the larger material and conceptual context that I have laid out. And it is to the issues of postwar abundance, resource scarcity, and system collapse that many energy crisis narratives looked when claiming crisis. We now turn to those early claims and some of the ways they made sense of disparate events as signals of creeping energy scarcity.

Assembling the Energy Crisis

In public discourse, the 1970s energy crisis is often associated with the 1973 oil embargo and the 1979 Iranian Revolution, the decade's so-called oil shocks.[143] The language of "shock" implies that the crisis struck suddenly, an unexpected jolt caused by forces largely external to the United States. Yet the fact is that claims of crisis emerged slowly in the years leading up to the oil embargo as experts tried to make sense of a disparate series of events that included electricity brownouts, fuel oil shortages, natural gas shortages, and peaking domestic oil production, as well as rising energy consumption and oil imports. A wave of books, reports, magazine articles, campaigns, and moralizing editorials warned of an impending or a hidden crisis well before the first oil shock. In a context of abstract energy consumption, in which most consumers procured gasoline, electricity, and fuel oil more easily than at any time in human history, the circulation of narratives, forecasts, facts, and figures pointing to growing energy demand and dwindling reserves helped to construct the energy crisis.[144] Often, these early narratives drew on notions of postwar mass consumption, neo-Malthusian anxiety, and Cold War ecological concerns about energy flow through systems to understand what was happening. When widespread oil scarcities and rising prices struck in late 1973, simmering fears congealed as a national crisis of energy.

In the booming 1960s, most energy experts foresaw a future of energy stability and abundance. Buoyed by faith in human ingenuity and capitalism's ability to unlock nature's bounty, energy forecasters acknowledged that finite resources must eventually be exhausted but argued that nuclear or solar power would eventually take over, just as oil had replaced wood and coal. RFF's 1963 *Resources in America's Future* study forecasted minor regional shortages but claimed that America would not have serious energy problems for at least forty years.[145] A Texaco forecast similarly concluded: "When these natural bounties approach exhaustion, man's ingenuity will rise to the occasion and these bounties will hardly be missed any more than we today miss the chopping of wood, the stoking of coal furnaces and the sifting of coal ashes which a great many of us can remember first hand."[146] Some experts expressed ambivalence about the sustainability of American wealth, but they retained enough faith in technological progress that they did not fear a serious crisis of energy.

M. King Hubbert's postwar surveys blended this modernist faith in technological development with the emerging environmentalist pessimism about the energy future. Early on, Hubbert worried about growing energy consumption driving rapid population growth and ecological degradation more than he worried about fossil fuel exhaustion, as is commonly assumed.[147] He adopted what he called a long-term "time perspective" on fossil energy, which meant using a timescale of thousands of years to highlight the relative transience of fossil fuels compared to the geologic and human record.[148] By its nature, the world's supply of oil would eventually run dry, though when that would happen was unclear. He urged undertaking a transition away from fossil fuels primarily to reduce destructive forms of growth.[149] But even Hubbert abstained from the language of crisis. Contemporary peak oil theorists and activists typically view Hubbert as a lonely, clear-eyed prophet who foresaw America's energy crises and their fallout long before anyone else.[150] Yet until the energy crisis, most of his public work on oil's approaching senescence expressed tempered hope that nuclear energy would replace petroleum, leaving overpopulation as the most pressing issue of the future.[151] He noted in 1956 that nuclear energy showed "promise" to supply enough energy for "at least the next few centuries," barring nuclear war or overpopulation. He illustrated this point with graphs that have not circulated nearly as much as his more famous "Hubbert's peak" graph, likely because they show nuclear power supplying significant energy inputs well into the future (fig. 1).[152]

Indeed, Hubbert's subdued enthusiasm for nuclear power appeared consistently in his writings about energy futures into the early 1970s.[153] Despite feeling pessimistic about economic growth and oil futures, he hesitated to declare a present or imminent energy crisis of potentially catastrophic proportions.

As was the case for many energy experts, public discourse in the United States did not expect an energy crisis as the 1970s began. Other environmental and social issues dominated the pages of magazines of mainstream opinion. *Life*'s massive double issue forecasting the 1970s fretted about overpopulation, overcrowding, and wilderness protection without any reference to energy consumption.[154] In the *Saturday Review*, philosopher Robert Heilbroner advised wise technological development and social equality as priorities for the new decade.[155] *The Nation* eyed the war in Vietnam, decolonization, and Japan's economic ascendency as the decade's defining issues.[156] Key voices in American capitalism also expressed an interest in environmental affairs, but no concern about energy scarcity. *Fortune* declared the environment a "national mission for the seventies," but it stressed pollution abatement.[157] *Forbes*, on the other hand, obsessed over the profitability of oil, focusing its forecasts on the uncertainties of international oil politics and domestic oil quota legislation. Instead of concern about scarcity, it espoused the view of America's business elite that growing oil demand would benefit the industry.[158] Texaco boasted that it was "looking forward to participating fully" in the coming energy bonanza.

Such confidence would be short-lived. Experts and oil companies soon began talking about an energy crisis as they confronted changes in America's oil production landscape, electricity brownouts, heating oil shortages, and new environmental legislation. These early diagnoses often conflated different forms of energy into a unitary category of "energy" in crisis in keeping with the postwar notion of America's fragile energy system. The prospect of oil scarcity primarily drove crisis narratives, but early crisis claims always exceeded oil as well. Fears

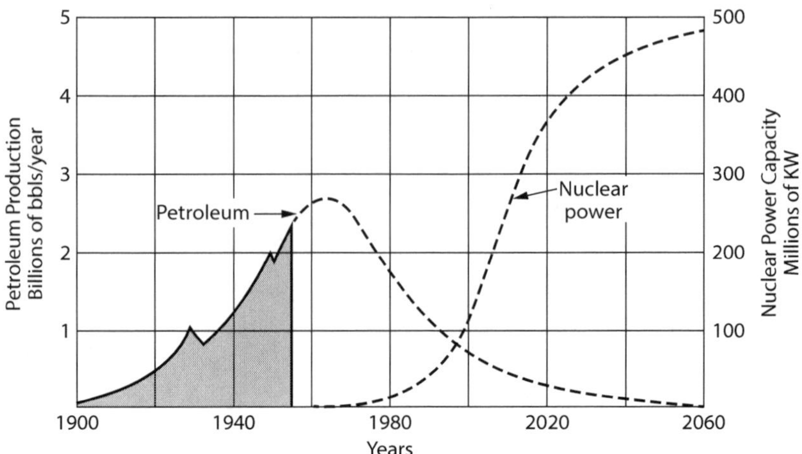

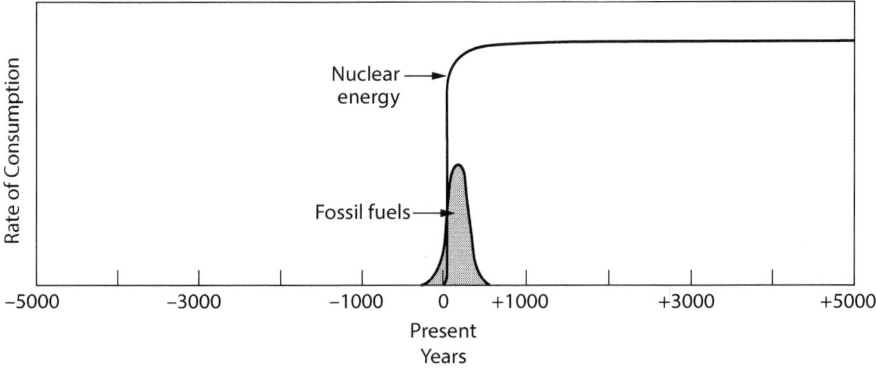

Figure 1. M. King Hubbert foresaw the end of the "Fossil Fuel Age" but was also initially optimistic about the potential for a nuclear future. Source: Hubbert, *Nuclear Energy and the Fossil Fuels*, 36.

about its insufficiency inspired calculation and forecasting about the other forms of energy that might replace it; as they were found wanting, the sense of crisis intensified.

American oil production reached a peak in 1970 at just over 9.6 million barrels per day.[159] Although, as Huber points out, this development was not immediately understood, it set the stage for the rapid growth of oil imports and evaporation of spare production capacity in the United States in 1972, both of which did inspire handwringing about the future.[160] In the midst of these momentous changes, energy shortages began to appear in 1970 in the form of brownouts and heating oil shortages. Concern about heating oil shortages in northeastern states first arose in the winter of 1970, prompting supply allocations from the Department of Defense and the Oil Import Appeals Board.[161] Major chemical and steel companies also suffered shortages throughout the winter, and the federal Office of Emergency Preparedness worried about potential power shortages in major population centers.[162] During the following spring and summer, media outlets began reporting fuel and power shortages and the potential for widespread brownouts and blackouts in the summer of 1970, some of which did occur. More predictions of power trouble appeared in fall 1970 and were followed again by more actual electricity shortages in early 1971. This pattern continued intermittently until the 1973 oil embargo and was bolstered by regional natural gas shortages as well.[163] Commentators often blamed new environmental legislation, such as the Clean Air Act, for the shortages, noting that smog controls forced utilities to substitute fuel oil for coal.[164] Scholars also debate the roots of these shortages, pointing to a variety of factors from price controls to quotas and the weather.[165] Regardless of the cause, a sense of impending crisis and frustration was gaining momentum as the world's richest nation found itself beset by sporadic shortages of energy and leaders unable to fix the problem. Energy consumption continued to grow, increasing by 5 percent in 1972.

The Nixon administration was reluctant to declare an energy crisis, though the increasingly frequent interruptions in the nation's power flow suggested to some bureaucrats that energy issues were becoming urgent. Nixon had created a cabinet task force on oil import controls in 1969 led by Secretary of Labor George Shultz, but it focused on the inefficiencies of Eisenhower's oil quota system. The administration was concerned with pricing oil to stabilize demand, rather than outright shortage in the present. Under Shultz's leadership, the task force argued against the quota system because it kept the price of domestic crude oil too high and exacerbated inflation.[166] Oil concerns did not abate, so in June 1971, Nixon delivered the first major presidential address to Congress on energy resources, in which he cited growing energy demand and new environmental laws as the most important factors in *potential* future energy problems. It was a crisis of flow from insufficient supply. Nixon's solution therefore included opening more federal lands to oil and gas exploration, nuclear power, clean energy research, and market-based energy conservation.[167] As part of his broader effort to combat inflation through wage and price controls, it also included complex oil price controls that did nothing to stall consumption.[168] Nixon avoided the language of crisis, and he seemed unconcerned about long-term energy consumption, concluding that America could have "both a high-energy civilization and a beautiful and healthy environment."[169] The resources were there; they just had to be exploited rationally.

Nixon's supply-side reasoning resonated with the oil industry, which always insisted that it could provide abundance to American consumers in the right regulatory environment and deeply resented new environmental controls.[170] The nation's postwar energy binge did not have to end if the country would loose its carbon capitalists on the land. Indeed, to achieve this goal, the industry played a central role in the articulation of an energy crisis early in the 1970s. Oil companies produced forecasts that overestimated future demand for oil and conducted a media campaign to posit that the impending crisis could only be averted by relaxing environmental legislation and opening more areas to drilling.[171] In the first instance, as Mitchell notes, industry forecasts argued for higher prices and greater access to land and resources by predicting high levels of future demand and stressing the high capital cost of development of new energy resources. A 1971 National Petroleum Council forecast argued that "in the long run, all indigenous energy supplies that can be developed will be needed," based, in part, on its projection of 4.2 percent energy demand growth per year until 1985.[172] It singled out federal land leasing, environment, health and safety, and tax policies as inhibiting the industry from supplying American consumers with sufficient energy from secure domestic sources. It warned that doubling domestic energy demand by 1985 meant more imported energy, unless the government changed its position on issues dividing corporate interests from workers and environmentalists.

Oil companies also discerned in the developing energy crisis a chance to reshape their image and tar environmentalists. The industry felt embattled by the burgeoning environmental movement and the 1969 Santa Barbara oil spill, which, according to the American Petroleum Institute (API) president, shifted "the national temper" and threatened negative legislative consequences in tax, import, conservation, and water pollution policy.[173] Under pressure from oil companies, the API moved its headquarters to Washington, DC, to heighten its lobbying and public relations power. From there, the advocacy group executed an advertising and public relations campaign designed to articulate "the importance of energy to the economy, to national security and to people's lives and to create an awareness of what the private enterprise U.S. petroleum industry was doing to ensure a continuing source of energy supply."[174] The institute spent nearly $10 million on advertising in 1971–72 to spread the notion of a looming "energy gap" between supply and demand, which became part of the lexicon of energy crisis discourse.[175]

The API and other industry representatives addressed the public and media frequently in 1972 and 1973 to confirm an impending energy crisis and to blame it on environmental legislation. In a press release from its annual meeting in November 1972, the API demanded that environmentalists weaken their demands in order to avoid a crisis. Continental Oil's executive vice president also took to the media to chastise "spoiled rotten" Americans who expected cheap energy but refused to open the Outer Continental Shelf or "let market forces set real values."[176] Print advertisements invoked a precarious future, asking if America could "head off energy shortages" by changing its policies in time.[177] During the 1973 oil embargo, oil industry executives and representatives called for rationing and offered advice about energy conservation.[178] One journalist noted the irony of industry support for an energy crisis and conservation narrative: "EPA [Environmental Protection Agency] and the oil giants should be the best of friends. Both want us to stop driving so much."[179] But for the oil industry, claiming a crisis was a way to discredit environmentalism.

Environmentalism and ecology also informed early claims of an energy crisis. Ecology provided what H. T. Odum called a "macroscope" through which to see the larger contexts for human energy development and exploitation. Consequently, sporadic energy shortages could appear as more than artifacts of politics or mismanagement;[180] they could be understood as feedback from systems being pushed to their limits. In July 1971, the *New York Times* published a three-part series on the "Nation's Energy Crisis," making it one of the first major publications to use the term. The series, written by scientist-journalist John Noble Wilford, explained the crisis as an artifact of a "bewildering" dilemma: high-tech, industrial civilizations use a lot of energy but demand clean energy to maximize quality of life. As demand rises and restrictions on producers proliferate, shortages ensue. A clean energy system will struggle to provide perpetual abundance.[181] Wilford considered nuclear power as a way to maintain wealth and a healthier environment, but he also framed the crisis as a challenge to the very possibility of continuous economic growth and asked whether growth was necessity for a reasonable standard of living.[182] By batting around ideas about halting growth, the nature of the biosphere, and Buckminster Fuller's concept of "spaceship earth," Wilford's series revealed the influence of ecological discourse on the development of the energy crisis assemblage. Sporadic shortages were rooted in environmental and sustainability issues, rather than merely the need to produce more and more energy, as the oil companies claimed.

The legacy of postwar ecology's conception of energy also informed early crisis claims. In 1971, the Sierra Club published a book coauthored by physicist John Holdren (later an adviser to President Barack Obama) and *Time* journalist Philip Herrera, who argued that the United States was already in the middle of an energy crisis "characterized by numerous, interrelated, unanswered questions about the future of America's environment, resources, technology, economy, national priorities, and policy."[183] They excoriated the legacy of American consumerism and its commitment to economic growth, but they also mobilized the language of systems ecology in noting how humanity's use of energy had surpassed its "metabolic requirements."[184] As experts would do throughout the 1970s, Holdren and Herrera affirmed the coherence of the US energy system as a space of crisis by devoting chapters to each form of energy in the system to discuss its nature, prospects, and drawbacks. This format became something of a genre of energy writing that is still with us, in which authors survey the technical history and futures of each energy resource in turn, from oil and coal to natural gas and solar energy, before speculating on potential future panaceas. Other popularizers of the energy crisis narrative, Lawrence Rocks and Richard P. Runyon, drew on postwar ecology and environmentalism to diagnose America as having an "accelerating" crisis rooted in its fundamental incompatibility with the ecological relations necessary to sustain life.[185] Rocks and Runyon linked the crisis to Malthusian overconsumption and overpopulation and provided a "timetable for an energy crisis" that predicted gasoline rationing, economic depressions, military conflicts over energy resources, water shortages, further overpopulation, and eventually, "massive and unpredictable environmental consequences" from global warming.[186] In this and other works, neo-Malthusian concern combined with a clear sense of a failing energy system and apocalyptic imagination to produce early energy crisis discourses conflating all shortages, from electricity to oil, to a generalized crisis of energy.

Ecological analyses of energy furthermore shifted the temporality of early energy crisis debates by locating a generalized notion of "energy" on much larger and longer scales. They discussed energy issues in terms of the interaction of multiple systems at varying scales over evolutionarily significant amounts of time. They intervened in public debate by introducing a sense of a deeper context from which to decide where to drill for oil and whether to develop nuclear power. In an introductory piece to a September 1970 special issue of *Scientific American*, ecologist G. Evelyn Hutchison noted that human economic activity had reached a scale that threatened the biosphere's long-term viability.[187] Other articles on natural energy cycles and human energy production belabored the broader point that natural systems are sensitive to large-scale human activities.[188] One year later, the magazine devoted an issue to "energy and power," introducing the solar ecology of energy to a wide readership. Articles echoed the earlier theme of earth's fragility but focused in detail on the myriad ways energy supports terrestrial life in relation to the biosphere, information, and different kinds of human societies. Energy was the key not only to human survival but to life itself; and it had to be managed with great care. One writer marveled at energy's "transcendent quality" as he sought to demonstrate the universal laws, and the limits, that governed human energy use.[189] Collectively, the issue suggested that continuing to violate the physical laws that governed energy would lead to disaster.[190] As the anthropologist Roy Rappaport concluded: "The general ecological perspective . . . suggests that some aspects of what we have called progress or social evolution may be maladaptive. We may ask if worldwide human organization can persist and elaborate itself at the expense of decreasing the stability of its own ecological foundations."[191] *Scientific American*'s survey of energy affairs expressed anxieties about the future of industrial civilization that had begun to emerge in the ecological work of the 1960s and raised the stakes of energy policy from the management of commodities to the possibility of restructuring American society. Although it resisted making an outright claim of crisis, it helped to establish the outlines of later energy crisis narratives.

Early claims of crisis also revealed the anticipatory nature of the energy crisis. It was often a crisis not merely of present problems, but of future disaster. The *Bulletin of the Atomic Scientists* was one of the first major scientific publications to elevate America's energy problems from shortage to crisis, in keeping with its origin story as a warning about the potential for nuclear disaster.[192] Its first issue on the energy crisis drew on postwar neo-Malthusianism and anticipatory rhetoric and assumed that a coherent crisis already existed as an object in need of analysis "from scientific, technical and economic points of view." Unlike *Scientific American*'s stress on limits and fragility, the *Bulletin* framed the crisis "as a crisis of uncertainties" about "the future of man and of his numbers; about his rational use of the planet for the perpetuation of his species."[193] The postwar population boom and rising expectations of living standards had intensified energy demand and spawned environmental concern. These anxieties, combined with sporadic energy shortages and the stubborn notion that economic growth required lavish flows of energy, generated uncertainty and a sense of looming crisis. Physicist Bernard Spinrad best expressed the telescoping of this dilemma by characterizing the energy crisis as "an *impending* energy crisis."[194] The language of a crisis that is both present and future was common, and it suggested the fundamentally anticipatory nature of the nascent energy crisis. Present shortages portended greater hardships to come. The articulation of that impending reality, which these early analyses initiated, insisted on developing solutions based on extrapolations about the future. For *Bulletin* writers, this often meant advocating nuclear power as the answer to the oil shortages and insatiable electricity demand to come. In its earliest stages, the energy crisis was a crisis of deciding how to preempt an extrapolated future. And that future, from the ecological view, threatened disaster. In his iconoclastic book *Environment, Power, and Society* (1971), H. T. Odum warned that cheap fossil fuels had caused human systems to become too dense, complex, and abstract, too fast. The United States achieved unprecedented energetic capacity to destroy the systems that support life.[195] Odum worried that the cultures like that of the United States, incubated in these energetic systems, would be unable to cope with an inevitable decline to "a minor energetic position in the earth system without a collapse and extinction of culture as we know it."[196] Odum appealed to a post–energy collapse future where "great gaunt towers of nuclear energy installations, oil drilling, and urban cluster will stand empty in the wind for lack of enough fuel technology to keep them running."[197] He hoped that such anticipation might create the political will to manage an energy transition preemptively, rather than become its victim. Many experts after him analyzed the crisis in a similar way, driven by similar hopes.

The Experience of Crisis

By winter 1973, experts had been talking about the energy crisis for nearly three years of sporadic shortages. Talk intensified and became more mainstream in January 1973 when *Newsweek* announced the arrival of the energy crisis (fig. 2). Then, in a widely discussed April 1973 *Foreign Affairs* article, State Department energy expert James E. Akins warned that the "oil crisis" was now a serious reality. Akins conflated oil with energy, even as he noted that influential oil forecasts had also done the same with oil and natural gas. He blamed the crisis, though, on the inadequacy of present and future global oil reserves, OPEC demands for autonomy and power, and the likelihood of consumer cooperation—all political rather than ecological or economic causes—to show that the United States had to "move on a variety of fronts" in the name of security and economic vitality, an echo of Cold War concerns about securing America's energy system.[198] Faced with accumulating signs of trouble, President Nixon finally acknowledged the crisis in an April 1973 address to Congress, where he anticipated that "occasional energy shortages and some increase in energy prices" were likely in the years ahead.[199]

Figure 2. Newsweek announced the energy crisis with an image of electricity, suggesting the conflation of energy forms inherent in the crisis. Source: *Newsweek*, January 22, 1973.

The arrival of the OAPEC oil embargo in October 1973, however, manifested the abstract projections and warnings of the early seventies in everyday experience in a material way. Prices jumped overnight, many stations ran out of gasoline, and gas lines began to form in November and December. The energy crisis assemblage congealed when the ecological and anticipatory discourses of crisis collided with the material and affective experience of scarcity in the winter of 1973–74. The impending crisis, which had been sporadically experienced in the form of natural gas, heating oil, and electricity shortages for years, now felt present for many Americans.

In late 1973, high prices and gas station lines became the primary symbols and experience of the energy crisis and its impact on American life.[200] At other times, light bulbs, powerlines, and chilly homeowners symbolized the looming crisis, but the arrival of the embargo shifted that emphasis toward prices and the impact of oil scarcity on automobility. Indeed, most people experienced the oil embargo by paying more for gasoline or waiting in line at overcrowded service stations. Gasoline, which had only risen seven cents per gallon between 1953 and 1972, suddenly shot up sixteen cents per gallon over 1973–74 and continued to rise. Lineups ensued as supplies fell short due to reduced imports and panic buying. Thus, the embargo "brought the shortage home to everybody," claimed the *Washington Post*, using language implying that the crisis had shifted from a subterranean and forecasted crisis into a public event.[201] The crisis was not felt until one had to pay up or wait to fill up, but it had been there for some time. Reduced oil imports, government-mandated rationing and shorter gas station hours, price gouging, and panic buying combined to create the maligned gas line phenomena that exacerbated public frustration. Media accounts related the anxiety, frustration, uncertainty, and acrimony created by searching for open service stations and waiting in line for hours for meager amounts of expensive fuel. In the winter of 1973–74, the experience and feeling of oil scarcity brought the crisis home to roost.[202]

Much of the frustration about the gas line revolved around its relationship to the temporal rhythms of postwar life. The gasoline line embodied the interruption of American energy flows while literally interrupting the postwar experience of time for millions of Americans. Many found themselves unable to get enough gas to maintain the petroleum-powered mobility to which they had become accustomed. Worse, they had to *wait* to receive their meager share. The perpetual movement of carbon capitalism and its concomitant experience of time ground to a halt. This idleness both stretched out the passage of time as boredom and sped it up in the form of the opportunity costs that a gas line exacted. Waiting for gas meant that you were not making or spending money. Gas lines were slow and boring and did not respond to individual needs or desires. One account described bored drivers drumming impatiently on steering wheels and "watching the sleet form patterns on their windshields."[203] Many detailed just how long people waited: thirty minutes, an hour, and sometimes much longer. The gas line both literally and figuratively undercut the accelerated mobility that had been a staple of the postwar economic boom, rendering time lost a central feature of crisis experience.

Beyond the boredom of waiting, the outrage of the gas line—invoked by populist paragon Howard Beale's "I'm mad as hell" diatribe in Sidney Lumet's *Network* (1976)—stemmed from the economic losses that it produced and the leisure time that it consumed.[204] Media accounts focused on how waiting for gas hurt the livelihood of those who relied on fossil-fueled automobility because they no longer had sufficient energy to turn their time into money.[205] Other descriptions and interviews with angry drivers focused on lost leisure and family time, now given over to waiting in line for gas to go to work. One man told the *Washington Post* that he brought his wife and sons with him to wait in line for gas because he had two jobs and little time for family life.[206] Thus, the energy crisis for many Americans manifested in the gas lines disrupting the swift flow of time, bodies, goods, and money that oil abundance had established as a norm in postwar America. Anxiety, frustration, and anger bubbled over in their inertness as the rhythms of postwar mass consumption and their purported affective states of freedom, happiness, and mobility were undercut by the slowing drip of the oil spigot.

The expectation of gasoline, and its necessity in a sprawling oil culture, combined with the frustration of waiting to generate feelings of anger, suspicion, and violence. As Finis Dunaway has noted, gas line imagery often framed the crisis as solely a question of supply and represented people waiting in line as frustrated, angry, and entitled individual consumers, rather than citizens with the capacity for collective action.[207] This framing comports with the Nixon administration's supply-side view of the crisis as a scarcity of inputs. Media reports stressed the anger and violence apparent in gas lines, which some excused as justified outrage at government incompetence for allowing the crisis to happen, or oil industry complicity in manufacturing the crisis to benefit from higher prices.[208] Gas lines became a stage for the display of what *Newsday* called "the varying emotions of hostility, jealousy, greed, and frustration."[209] Drawing on the legacy of postwar consumerism and populist rhetoric, columnist Tom Wicker opined: "If there is blame, and there certainly is, it lies only marginally on the hapless driver of the great American gas-guzzler or the housewife-consumer of electricity, both victims of relentless advertising, and neither of whom failed to build sufficient refinery capacity when it was obviously needed . . . or lobbied for oil import quotas . . . or gets a depletion allowance to help explore for more gas."[210] Blame had to "start at the top." The crisis was systemic. Satirist Russell Baker penned a State of the Union address while sitting in a gasoline line that excoriated the president, Congress, and "affiliated big shots" for their insulation from the mundane frustrations that plagued average American lives.[211] The experience of the gas line manifested the crisis for many people and enflamed the populist distrust of elites that marked much of the decade. It also, as the Wicker quotation suggests, often drew on the postwar formation of a consumer society and broader debates about whether scarcity was a real or manufactured threat. Yet once the lines dissolved, public memory tended to refer to them as a quaint, common experience that was now firmly part of the past.[212] The return of gas lines in 1979, along with anger, violence, frustration, and suspicion, revived public discourse about the energy crisis.

Conclusion

To claim a crisis is to make a political claim rooted in a modernist milieu. This chapter has explored the assembling of the energy crisis in support of its thesis that a crisis of energy could only have occurred in the early 1970s. The energy crisis was a historically specific assemblage that brought together key material and conceptual developments from the postwar period with the experience and narration of energy scarcity in the early seventies. Although early crisis narratives derived from different perspectives and aims while the experience of scarcity was ambivalent at best—surprising, maddening, disorienting—they often referred to the materiality of postwar consumer culture, neo-Malthusian fears of scarcity, and the calculative tools of ecology to understand energy as an abstract input in systems that had suddenly become unreliable.

Once claimed, crises also beg interpretation.[213] To understand how the energy crisis supported the neoliberal shift in US political culture, the next chapter explores the meanings that different constituencies attributed to the energy crisis as citizens, leaders, and experts tried to make sense of such unprecedented peacetime scarcity and to chart a path through the future. The realization of any future, according to most interpretations, required preemptive action in the present to manage the nascent transition to a new age. Ecologists, as noted, were central to these interpretations of what the energy crisis meant for the future, and to the reinterpretation of the past as a golden age of cheap energy now passing away.

Chapter Two

"A Time to Choose"

Interpreting the Energy Crisis

We look to the future with pleasure
we need no fossil fuel
get power within
grow strong on less
—GARY SNYDER, POET, 1974

Vigorous programs should be quite successful in finding oil. . . . The north slope of Alaska may be as rich in oil as some parts of the Persian Gulf.
—HERMAN KAHN, FORECASTER, 1974

The energy crisis fueled a debate over its meaning that sustained it for years after the 1973–74 oil embargo ended. Although claims of crisis drew from the experience of energy scarcity and economic inflation that many Americans shared during the embargo, the crisis itself primarily unfolded and evolved through a swirl of narratives about what those experiences portended. Competing narratives aiming to explain what was happening and what it meant for America thus constituted the crisis for most of the decade.

Energy crisis narratives derived much of their constitutive import from the subterranean nature of oil. For most people, their use of oil products was and is heavily mediated by abstract extraction and delivery systems that bring oil from the ground to the gas tank. Few people ever see crude oil or even gasoline unless there is a spill, and even then, most spills go unseen. We can certainly see and feel the capacities that fossil fuel consumption endows, but it is beyond our senses to know the status of the raw resource from which we derive our godlike capacities. Expert claims, forecasts, and other forms of discursive mediation therefore contextualize the experience of energy scarcity and shape popular knowledge about fossil fuel reserves. A nineteenth-century farmer could point to a deforested landscape to explain his struggle to heat hearth and home, but a mid-century suburbanite must rely on expert claims about oil reserves and futures to understand her struggle to fill her gas tank and pay her energy bills.

Although there were many interpretations of the energy crisis, mainstream debate revolved around two core positions, as embodied by two early 1980s texts: *The Global 2000 Report* (1981) and *The Resourceful Earth* (1984). The former evolved out of Jimmy Carter's framing of the crisis as an energy supply and environmental policy issue, and from his concern that the federal government lacked the capacity to forecast and manage environmental problems. He had been deeply disturbed by the energy crisis and by the Club of Rome's stark neo-Malthusian prediction of precipitous economic and demographic decline.[1] So, soon after taking office in 1977, Carter commissioned the Department of State and the Council on Environmental Quality to write *The Global 2000 Report* on "probable changes" to global population, natural resources, and environmental health to the year 2000 to aid government policy and planning.[2] Extrapolating from present population and resource trends, the report predicted a world "more crowded, more polluted, less stable ecologically and more vulnerable to disruption than the world we live in now" by the year 2000. Its "reconnaissance of the future" argued that continued economic growth threatened to push global consumption beyond the earth's carrying capacity, with particularly troubling consequences for the world's poor.[3] It warned that only a swift international response, led by the United States, could avert the catastrophic collapse of global ecology and economy.[4] *Global 2000*'s population and resource futures echoed many 1970s environmentalists and found a large audience.[5]

Many conservatives, neoliberals, and think tank experts were deeply disturbed by Jimmy Carter's views and the prospect of more government planning. In 1984, the Manhattan Institute, economist Julian Simon of the Heritage Foundation, and futurist Herman Kahn of the Hudson Institute published *The Resourceful Earth: A Response to Global 2000* to show how "dead wrong" *Global 2000* was.[6] Kahn, who achieved infamy as a forecaster of the "unthinkable" in his 1960 book, had become Washington's most influential futurologist in the 1960s and 1970s.[7] Contra most environmentalists, he foresaw a "hyper-industrialized" future of abundance in which people were more likely to be bored than cold and hungry. Kahn died before the publication of *The Resourceful Earth*, but it represented a culmination and continuation of his ideas. The book rejected the limits to growth, arguing "that the nature of the physical world permits continued improvement in humankind's economic lot in the long run, indefinitely."[8] Material progress could only be stalled by the assumption that nature is limited and by regulatory efforts that such an assumption supported. Simon and Kahn claimed that the earth's vast resources had virtually endless productive potential that human ingenuity—

structured by competition in a free market—could unleash. *The Resourceful Earth* downplayed the energy crisis as a "bogeyman" and insisted that the availability of useful energy would continue to increase "until our sun ceases to shine in perhaps 7 billion years."[9]

The Resourceful Earth and *Global 2000* shared little common ground. Where *Global 2000* imagined a dystopian, overpopulated future of scarcity, poverty, and environmental ruin, *The Resourceful Earth* projected a healthy and stable world of leisurely abundance.[10] *Global 2000* foresaw a future of limits; *Resourceful Earth* fixed on untapped possibility. One advised direct state intervention to reduce population, production, and consumption while the other favored the free market to allocate resources and incentivize production.[11] Despite these differences, however, both texts challenged America's Keynesian liberal consensus as outmoded and ultimately destructive. They represented the culmination of the seventies struggle to come to terms with the meaning of the energy crisis, to understand where it came from and what it portended.

This chapter explores the interpretative struggle over the nature and meaning of the energy crisis that informed *Global 2000* and *The Resourceful Earth*. Why focus on perceptions of the crisis rather than law or policy? The nature of the energy crisis was less a matter of brief shortfalls of oil imports and rising prices than it was of the potential futures that these events embodied and the stories that experts told to contextualize them. This narrative swirl was diverse, but it essentially contained two rival challenges to Keynesian liberalism, which I am calling the environmentalist and market narratives of crisis. The environmentalist critique disavowed the excesses of postwar capitalism and its regime of mass consumption as hostile to the viability of life on earth, while market advocates viewed the same period as a time of excessive government intervention that had spawned overconsumption and anemic production. Both narratives thus warned of future energy scarcity based on competing notions of the economy of nature and the nature of the crisis, and both articulated competing paths to the future. Crucially, their competing mobilizations of the energy crisis demonstrate the political flexibility of crisis, which in this period undermined Keynesian orthodoxies about how to manage the economy.

There are five sections to this chapter. After sketching the main contours of postwar Keynesian liberalism, I divide the mass of energy crisis narratives into three temporal categories tied to the past, present, and future. The first explores historical narratives that laid the blame for the energy crisis at the feet of American citizens and institutions. The second notes how the present became a time of transition when viewed in light of the energy crisis. The third considers the mobilization of competing notions of the future that experts divined in their analyses of the energy crisis. This organization aims to evoke the cacophony of voices pronouncing on the energy crisis while also bringing to the fore the dominant market and environmentalist positions that shaped how the energy crisis would be viewed and acted on in the future. The chapter concludes by reflecting on the critique of "crisis" that this historical analysis embodies.

What Was Keynesian Liberalism?

Historians have proposed a variety of names to characterize the period of economic prosperity in American life from roughly the end of the Second World War to the beginning of the energy crisis. Whether named the "great exception," the "great compression," or some other designation, the postwar period was a time of growing and more equitably distributed wealth in the United States. It was unprecedented in its time and has not since been matched. Between 1945 and 1973, for instance, home ownership grew by more than 20 percent. Private car ownership soared. Incomes rose for all income brackets. Unemployment frequently dipped below 5 percent and income inequality shrank.[12] The minimum wage nearly tracked with a living wage before falling far below it in the 1970s. Even political bipartisanship, argues Jefferson Cowie, rose to a level never seen before or since.[13] Poverty and inequality did not go away, but postwar America achieved an impressive degree of upward mobility and wealth distribution unmatched in most of the rest of the world.

Among elite experts in America's top universities, newsrooms, and government agencies, there existed a consensus that the nation owed much of this postwar success to an economic policy paradigm inspired by the British economist John Maynard Keynes. The so-called Keynesian liberal consensus was characterized by the core idea that experts in government could and should manage the economy to ensure steady and sustained economic growth, as well as full employment and stable prices.[14] Growth was a central aim, and government management had a key role to play in shepherding it. A key tool for doing so was deficit spending to stimulate demand that would maintain high production and employment levels.

This paradigm was particularly dominant in the Kennedy-Johnson years, when the so-called New Economists (Walter Heller, Gardner Ackley, and Arthur Okun) steered economic policy. Given the tremendous economic dynamism of the US economy in the 1960s, Kennedy's Keynesian liberals exuded confidence in their ability to manage economic affairs with hitherto undreamed of precision. Economist Paul Samuelson compared his fellows to "artists, mixing the colors of our palette" with monetary and fiscal policy to calibrate levels of taxation, spending, and money supply to produce full employment.[15] *Time* named Keynes its "Man of the Year" in 1965 amid years of low unemployment.[16] By 1971, even Republican President Richard Nixon claimed to be a "Keynesian in economics" after proposing a deficit budget to tackle the threat of rising unemployment and economic stagnation.[17]

Energy crisis debates were in many ways a referendum on this postwar liberal consensus, questioning its verities and calling for change, and they did so in a broader climate of crisis. Indeed, the energy crisis stalked a nation already beset by creeping doubts and recriminations, the boasting of the neo-Keynesians notwithstanding. The 1960s had been a decade of trenchant critique of American society and its institutions. The civil rights, women's rights, environmental, and anti-war movements all questioned the goodness of America and demanded changes to its social norms and values, and to its laws. The civil rights movement's demands for equal opportunity and equal protection of the law for African Americans were often met with violence that culminated in the riots of 1967 and 1968, leading a large majority to affirm Nixon's "law and order" campaign. The nation's military failure in Vietnam created unease about the limits of US power and fomented a sense that perhaps it was already in decline, a much shorter reign than that of the British Empire.[18] Moreover, although Americans in general were richer than they had ever been in the 1960s, "affluence was as much an ideology as a description of U.S. society" as the 1960s came to an end.[19] In 1970, public discourse rediscovered that 60 percent of working Americans either lived in poverty or found themselves hovering between poverty and what the Bureau of Labor Statistics deemed a modestly adequate income.[20] Thus, a sense of accumulating crisis formed the context in which experts and other observers articulated the three forms of energy crisis interpretation traced in this chapter. Both the market and the environmentalist narratives that are the focus of this chapter undermined the confidence of Keynesian liberalism. Embedded in the spirit of self-critique that marked the late 1960s and 1970s, these narratives blamed America for its energy woes and sought alternative paths forward.

Blaming America

Many experts first interpreted the energy crisis as a judgment against the nation's postwar mass consumption binge. The appearance and ongoing possibility of shortages in the massive flows of energy resources required to power American lives suggested a moral failing on the part of the nation to steward its resources wisely during a time of abundance. Whether that failure was inherent in the structure of industrial capitalism, which had ignored the laws of nature by living on borrowed natural capital (fossil fuels) rather than natural income (solar energy), or whether it was a failure to trust various American institutions to shape behavior and expectations, like the free market, depended on the interpreter. But most interpreters echoed the nation's Protestant heritage and ongoing evangelical resurgence, which were both ambivalent about consumption, by narrating the energy crisis as a moral failure to keep the American house in order.

Concern with moral judgment implied American responsibility for an energy crisis that many interpreted as a crisis of overconsumption.[21] It was a bitter but just recompense for the nation's wasteful habits. Analysis in newspapers and magazines routinely highlighted the disparity between American and European energy consumption and its relation to standards of living, pointing to a US citizenry that had grown decadent. *Business Week*'s environmental editor chastised America for becoming an unethical "throwaway society" characterized by a "frontier ethic" that was "astonishingly profligate [with] the earth's natural resources."[22] The *Washington Post* likewise declared the crisis to be one of "self-indulgence on a gigantic scale," while *Time* noted a popular view of the crisis as the aftermath of "a long Belshazzar's feast of energy gluttony . . . [in which] Americans were being called to a bitter reckoning."[23] Stewart Udall, the activist secretary of the interior of the 1960s, labeled Americans "energy pigs" for devouring more than their fair share of global energy resources.[24] For these observers, the energy crisis seemed yet another symptom of the "Me Decade" in which too many people narcissistically prioritized their own impulsive desires rather than those of the community or nation.[25]

Early representations of the energy crisis used images of necessities like heat and light to express the threat that shortages posed to American well-being, but after the oil embargo and its frustrating gasoline lines, the automobile became the symbol of sinful energy indulgence, a sign of American excess now coming home to roost. US car culture became for some a naughty or illicit affair nearing its end.[26] In December 1973, at the chilly nadir of the oil embargo, a *Time* cover declared "The End of an Affair" between American motorists and big cars.[27] The cover story labored to extend the metaphor, noting that Americans felt a "high level of passion" for "their huge, gleaming cars" that afforded social status, kinetic excitement, and freedom.[28] As if to mourn the passing of this passionate relationship, the article featured a brief photo essay about the pleasures that big cars made possible: camping in New England, romantic encounters, roadside restaurants, and loading up with food at the supermarket. Perhaps this affair with pleasurable waste, such articles implied, was the original sin at the root of present troubles.

The moral language of wastefulness frequently appeared in presidential diagnoses of the crisis. Presidents Richard Nixon and Gerald Ford used this critique reluctantly, speaking of a collective responsibility to abandon bad habits while pursuing programs of intensive energy production under the banner of "Project Independence." In a January 1974 speech, Nixon compared the nation to an addict in danger of backsliding "into the wasteful consumption of energy."[29] In 1977, Ford noted that the American "economy and style of life . . . built upon cheap and abundant energy" and buoyed by government regulation designed to keep energy cheap had created "wasteful and inefficient" use of natural gas and oil.[30] Most infamously, Jimmy Carter unabashedly rebuked Americans for their "wasteful habits."[31] In his first national address on energy, Carter proclaimed: "Ours is the most wasteful nation on Earth."[32] Failure to reverse the nation's descent into decadent overconsumption, he warned, would only exacerbate dependence, poverty, and environmental ruin.

The tendency to bemoan the energy crisis as an American failing coexisted with ethno-nationalist criticism of "Arabs" in the US media, which often featured images of Arab "oil sheiks" exploiting the addictive urges of American oil consumers.[33] But if we are to grasp the relationship between energy crisis discourse and neoliberalism, it is important to note that many commentators viewed OAPEC's demand as a reasonable action in a realpolitik power struggle for resources and economic power. It rendered the cartel more rational than the populist image of the volatile Arab "oil sheik" would suggest.[34] *Forbes* argued that "the Arabs have every right to use their economic power. But remember, their interests and ours will not always coincide."[35] The embargo involved one set of market actors seeking rationally to maximize gains, and Americans had better learn to compete.[36] A letter writer to *Time* mixed orientalist condescension with self-criticism when he noted: "It is obvious that the Arabs have come of age because they no longer rely on verbal barrage and senseless acts of terrorism to advance their interests. One cannot help admiring them for realizing that as the Russians understand force, the Americans, and to some extent the West Europeans, understand when they are hit in the pocket."[37] The energy crisis strengthened the anti-Arab xenophobia that underwrote US resolve to shape Mideast politics while at the same time inspiring soul-searching about US policies and values through which the market eventually emerged as an explanatory grid for the energy crisis.[38] As discussed later in this book, these two strands came together in the 1980s, when the US state sought to retake control over global oil pricing from OPEC using futures markets.

If the energy crisis signaled a judgment of History for US profligacy, decadence, and dependence, why did Americans find themselves in this mess? Two dominant explanations emerged, one emphasizing market dynamics and the other an unsustainable society. Many free market economists and think tanks argued that US government policies distorted market prices and encouraged wasteful consumption. In this interpretation, America had since the New Deal repudiated the source of its freedom and vitality and embraced various forms of dependence.[39] The American Enterprise Institute (AEI), a think tank established in the late 1930s in part to counter the effects of the New Deal, became very active in advancing a free market narrative of the energy crisis. The AEI formed a National Energy Project and published several reports and pamphlets about the crisis. It argued that the United States had gone astray with the New Deal by abandoning the free market, claiming in numerous publications that the market had not been sufficiently free to "signal the higher true cost of energy to the consumer."[40] Without free market price signals, the US economy relied on bureaucracies that, according to economist and Nixon adviser Herbert Stein, could not "possibly obtain the mass of complex, detailed, future-oriented information continuously needed for such regulation."[41] W. Philip Gramm, an economist and soon-to-be senator, claimed in the *Wall Street Journal*: "'Crisis' as opposed to simple scarcity, results from market disruptions," such as price controls on oil. America was running out of energy simply because its policies continued to "prevent the free market from working."[42] In the long term, right-leaning advocates hoped the market would boost production and restore supplies, but in the short term, they saw it as the best option to rein in consumption.

While market interpretations narrated the energy crisis as punishment for abandoning free market principles, environmentalist narratives framed the crisis as a judgment against the unsustainable and destructive economic system of the United States and its postwar consumer society. By embracing mass consumption, Americans had violated the laws of ecology and thermodynamics governing the energy to flows that supported life itself. Postwar industrial capitalism prodigiously wasted finite energy resources that would eventually run out. As EPA Administrator Russell E. Train told *Newsweek*, the United States wasted "a helluva lot" of energy.[43] The environmental think tank the Worldwatch Institute, established by environmentalist Lester Brown in 1974, calculated that in 1975 alone Americans wasted more fossil fuel than most of the rest of the world used in that year: a "staggering" example of profligacy rooted in capitalism's mythic pursuit of growth.[44] The shortages of the early 1970s, in this framework, were an early warning of a deeper disaster slowly gathering momentum.

This narrative's critique reflected the ethos of the larger environmental movement in the United States. Both environmentalism and ecology, which were often assumed to be ideological bedfellows by the time *Newsweek* declared the "Age of Ecology" in 1970, often positioned themselves as moral alternatives to the destructiveness of capitalism.[45] In this, they built on a longer legacy of conservationist and environmentalist critiques of overconsumption, from Henry David Thoreau's critique of conspicuous consumption to conservationist and wildlife biologist Aldo Leopold's notion of an "ecological conscience."[46] In the age of nuclear weapons, mass chemical use, and multiplying extinctions, the tenor of environmentalist critique intensified. The understudied writer and naturalist Peter Matthiessen framed white America's relationship with nature as exploitative and called for an ethical renewal. Rachel Carson's *Silent Spring* condemned the indiscriminate use of pesticides as both a scientific and a moral tragedy.[47] The moral sensibility that fueled environmentalism also motivated the first Earth Day on April 22, 1970, where activists linked their cause to larger structural wrongs. As organizer Denis Hayes declared, "Ecology is concern with the total system—not just the way it disposes of its garbage."[48] The environmental crisis did not emerge through a series of bad decisions or technologies but rather was rooted in an exploitative disposition toward nature in need of reformation. It expressed the structural wastefulness of US society.[49]

As the 1960s counterculture evolved to embrace environmental concerns, particularly in the back-to-the-land movement and various forms of neo-agrarianism, key figures extended its critique of mainstream American lifestyles and values to include energy waste as a manifestation of a widespread and unconscious acceptance of an unnatural and unhealthy way of life. Homesteaders and beat poets like Gary Snyder and Allen Ginsberg, as well as elements of the nascent New Age movement, abhorred centralized power, control, and conformity, traits that they ascribed to a high-energy culture that was inimical to their ideal of a decentralized and agrarian America. For them, the energy crisis seemed to be a judgment on America's inauthentic culture.[50] Snyder challenged the counterculture to confront the question of growth and turn for sustenance to internal and spiritual energies rather than the commodified energy forms (gasoline, electricity, etc.) that ultimately impoverished human life.[51] Deepening dependence on excessive external energy inputs, for these writers, prevented the formation of authentic community and self-understanding. Neo-agrarian writer Wendell Berry, for instance, wrote in 1979 that industrialism and its centralizing technologies eroded communities and coherent selves, ultimately alienating people from each other and the earth in the pursuit of more energy, power, and money.[52] Similarly, Snyder's Pulitzer prize–winning poetry collection, *Turtle Island* (1974), framed its exploration of "how to be" in a world of crisis around ten facts about the relationship between the environment and the American economy that featured H. T. Odum's notion of net energy and the West's "fossil fuel subsidy."[53] Energy forms shaped social forms and therefore had to be renavigated in the present to heal a sick society. Fossil fuel systems alienated people from the land and from themselves, but as Snyder said in 1979, small-scale, "decentralized energy technology could set [America] free" from its alienation and enslavement to excess. The centralizing, technocratic energy solutions that had encouraged the development of a rapacious consumer culture and created the crisis in the first place, were immoral and "unimaginative."[54]

It is perhaps unsurprising, given the prominence of such critiques, that the metaphor of "addiction" circulated widely to frame the unhealthy fossil fuel dependence that both the market and environmentalist narratives discerned in US consumer culture. The environmentalist narrative decried the nation's apparent inability to reduce fossil fuel use in spite of severe social and environmental consequences, as in Snyder's vision of a nation "addicted to heavy energy use, great gulps and injections of fossil fuel."[55] However, others largely abhorred dependence on *foreign* supplies of oil. Beginning in 1970, when domestic oil production failed to keep pace with demand, many commentators worried aloud about American dependence on foreign oil as a threat to national security.[56] The Nixon administration responded with Project Independence, an initiative to achieve energy independence by 1980 that drew on a deep cultural tradition of self-reliance.[57] Dependence on the resources of other nations was anathema for many middle-class Americans accustomed to a sense of global leadership; when aversion to it joined the growing sense of guilt and self-recrimination over energy profligacy, the image of oil addiction went mainstream, sparking interest in "self-reliant" forms of living as a way to regain independence through energy frugality.[58] The image of Uncle Sam as an oil addict and of Arabs as drug pushers appeared by the mid-1970s in newspaper editorials and cartoons. Although Natasha Zaretsky has expressed surprise at this criticism of American consumers, the moral condemnation of such imagery echoed critiques of the energy crisis as a sign of sickness, whether eco-logical, economic, or geopolitical.[59]

Historical judgment narratives of the energy crisis in either their "market" or "environmental" form dominated crisis debates throughout the 1970s. In addition to the cultural challenges of the 1960s, they also drew strength from the decade's economic challenges, which were linked to the energy crisis itself. Stagflation (the simultaneous persistence of inflation and stagnant economic output that be-deviled neo-Keynesians), slowing productivity growth, stalled wages, and a host of other maladies fueled a cultural mood of self-criticism that questioned the na-tion's moral standing. President Jimmy Carter channeled these energies in his July 15, 1979, "Crisis of Confidence" speech, one of the key moments in the political rise of neoliberalism. As Daniel Horowitz has recounted, Carter delayed his ad-dress and retreated to Camp David to confer with intellectuals about the best way to confront America's energy problems.[60] After weeks of speculation about the delay, Carter declared that the energy crisis had exposed a "moral and spiritual crisis," drawing on anxieties about the Me Decade and its pervasive "culture of narcissism."[61] He complained that "all the legislation in the world" could not "fix what's wrong with America" because Americans had traded traditional values and discipline for individualistic indulgence and gave no thought to tomorrow. Hence America's declining productivity, cultural emptiness, and uncertainty: a spiritual crisis to compound the energy crisis. Carter declared the energy crisis to be a turning point that offered a choice between self-interest and fragmentation or the "restoration of American values." Energy would be the "battlefield" on which the nation would test its resolve and renew itself.

Carter was righter than perhaps he knew. Energy was indeed a battlefield, but the language of renewal came from the right. Carter's diagnosis received a mixed response. According to one poll, three-quarters of Americans agreed that there was a very real "crisis of confidence" in the United States.[62] But as problems per-sisted, Carter's leadership came under increasing scrutiny, and pro-market com-mentators used the speech to question the government's ability to solve the crisis. Media reports noted Carter's "moralistic tones" and questioned his political fu-ture.[63] The *Wall Street Journal*'s retort was perhaps the most prophetic insofar as it vehemently resisted Carter's diagnosis of a spiritual crisis: "Some people like to party, and some like to teach Sunday School. What nobody likes is setting their thermostat at 68 degrees."[64]

This response anticipated the tenor of energy crisis discourse in the 1980 pres-idential election race, skillfully exploited by Ronald Reagan's candidacy declaration invoking the need for a "spiritual revival." But contra Carter, Reagan blamed the energy crisis for the spiritual crisis, rooted in government meddling and poor leadership. He rejected the argument that America's problems stemmed from its "past excesses" and promised to let the free market guide energy production and prices.[65] Self-critical discourses of sacrifice, ecological sensitivity, and moral judgment, as widespread as they had been in previous years, were beginning to lose their hold. In retrospect, Carter's speech was the apex of the moral judgment narrative, after which the pro-market narrative adopted a more optimistic tone in its advocacy of markets to solve the energy crisis. The Reaganite anticipation of a brighter future through the market and inversion of the relation between the energy crisis and America's other problems set the tone for public discourse in the 1980s.

Times of Transition

Today, awareness of the historicity of fossil fuels is commonplace among scholars. Energy history and the energy humanities are lively fields, and climate change has turned fossil fuel energy dependence into a problem that must be solved—a historical aberration with an expiration date.[66] But such awareness is not new. Fossil fuels have long been stalked by fears of exhaustion that ebb and flow over time in concert with other cultural anxieties. During the 1970s energy crisis, anxious accounts about fossil fuels differed from past concerns about peak oil or exhaustion in that they were historical narratives about the effect of fossil fuels on the shape of the human past that reframed the present as a time of transition. They were also far more widespread in expert circles and in mass media discourse. Thus, even as they celebrated the nation's bicentennial in the 1970s, experts and policymakers used the term "Fossil Fuel Age" to express the sense that finite energy sources had fueled the technological progress that propelled US exceptionalism and the nation's faith in its future, as well as the material pleasures of the postwar period and its dogged pursuit of economic growth. Adherents to the environmentalist critique sought to denaturalize the Fossil Fuel Age by narrating it as a time of exception to be left behind, while market narratives sought to naturalize its material abundance as a desirable product of the market. The combined effect of both narratives was to posit the present as a moment of transition away from the postwar Keynesian liberal order, figured as wanton and doomed to fade into the mists of the past, though not without anxiety and nostalgia for fossil-fueled pleasures that suddenly seemed ephemeral.

Rear Admiral H. G. Rickover, the "Father of the Nuclear Navy," first coined the term "Fossil Fuel Age" in a 1957 address titled "Energy Resources and Our Future."[67] Speaking to the Minnesota State Medical Association, Rickover recounted the history of human energy use from ancient times to the present in a historical analysis that connected surplus energy to higher standards of living and geopolitical power. Noting the dangers of US dependence on nonrenewable fossil fuels, he advocated conservation and technological development while predicting that America's next generation ought to prepare to "ring out the Fossil Fuel Age."[68] When Rickover used the term again in his 1972 testimony before Congress, he revealed that reading Truman's *Resources for the Future Report* (1952) had dashed his belief in America as a "country of limitless space and boundless resources." He came to understand instead that postwar America was "living high on resource capital which—no matter how large—was bound eventually to be exhausted."[69] The Fossil Fuel Age had been grand but could not last.

Rickover's notion of a transitory Fossil Fuel Age circulated in 1970s discourse to signify a newly minted historical period that was dying at the moment of its discovery. Indeed, dozens of periodizing narratives reimagined the present in light of a new awareness of humanity's energetic history borne of the energy crisis.[70] These narratives understood the present and the recent past as a transitional phase of fossil-fueled abundance between longer periods of more moderate solar energies in the past and a contested post-carbon future. Even though direct shortages affected the United States and Western Europe, and were artificially imposed by the oil embargo, such assessments spoke in a universal register about the history and energy future of "man." Their universalizing tone acknowledged the fact that the United States and Western Europe consumed vastly disproportionate amounts of the world's carbon energy resources but nevertheless insisted that the crisis was global and required a global response, much like Rickover initially did when he attributed mid-century poverty in Asia to energy poverty rooted in technological stagnation.[71]

Fossil Fuel Age narratives could be used to support a variety of different political initiatives. For instance, the Joint Committee on Atomic Energy (JCAE), established by Congress in 1946 to oversee the development of nuclear power in the United States, claimed in a 1973 report that the United States now found "itself in the 'twilight' of the fossil fuel age." The nation had exhausted its best carbon energy resources to build "a technical and industrial society the likes of which the world has never seen." Unfortunately, the impending end of the Fossil Fuel Age required the *world* to enter a new "'age' or 'era'" before it was too late. The JCAE recommended a period of conservation and intensified domestic carbon exploitation to buy time to develop alternative energies for the next age.[72] This declaration of a closing Fossil Fuel Age thus comported with the JCAE's vision of a nuclear-powered future. It stressed the density and complexity of the technological society that fossil fuels had enabled and the need for atomic power to sustain it. The designation of the present as a fading "age" helped to rationalize a period of even greater resource exploitation under the teleological understanding that it would soon come to an end as US technical expertise ushered in a post-carbon era of nuclear power. Indeed, these sorts of periodizations always relied on speculation: How much carbon energy is left, and where? How will demand grow? How long do we have to use it? When will it run out? The energy crisis helped to demarcate the present and recent past as a distinct age but gave less guidance about how to realize a new future. Note the committee's demand forecasts (fig. 3): by naturalizing the growth of energy demand, nuclear advocates happily bid the Fossil Fuel Age adieu, knowing that only nuclear power could meet the demand they predicted. The conflation inherent in the "energy crisis" and measures like "total energy" could be productive for a variety of constituencies.

Narratives such as this, in which the energy crisis marked the end of a universalized Fossil Fuel Age, marked the revival of the energy-based narratives of human history initially developed early in the Cold War by Leslie White and Fred Cottrell. One popular book from the seventies noted that it was important to reframe history as the story of "the relationship between energy use and the evolution, present functioning, and potential future of human society." It called this approach an "energetic viewpoint," wherein the entire scope of human activity is reduced to "energy-using processes depending entirely on the continued availability of energy resources."[73] Its universalizing re-narration of the past dovetailed with the universalizing narrative of the energy crisis—as it related to food, population, and pollution—that marked late contemporary studies and recommendations, like the *Global Future* follow-up to Carter's *Global 2000* report.[74] The energy crisis may have resulted from the overconsumption of the industrial West, but the narratives of human-energy interdependence that flowed out of it implicated the world in Western schemes for energy development and efficiency.[75]

The intuition that Americans were living through a transition away from a fast-fading Fossil Fuel Age also gave rise to the notion of the postwar period as a "Golden Age" in American history that was now under threat.[76] Despite their divergent timescales, the energy crises in both in the Golden Age and the Fossil Fuel Age clarified the recent past as a unique period of growth in population, wealth, and leisure driven by abundant fossil fuels.[77] Only fossil fuels could have

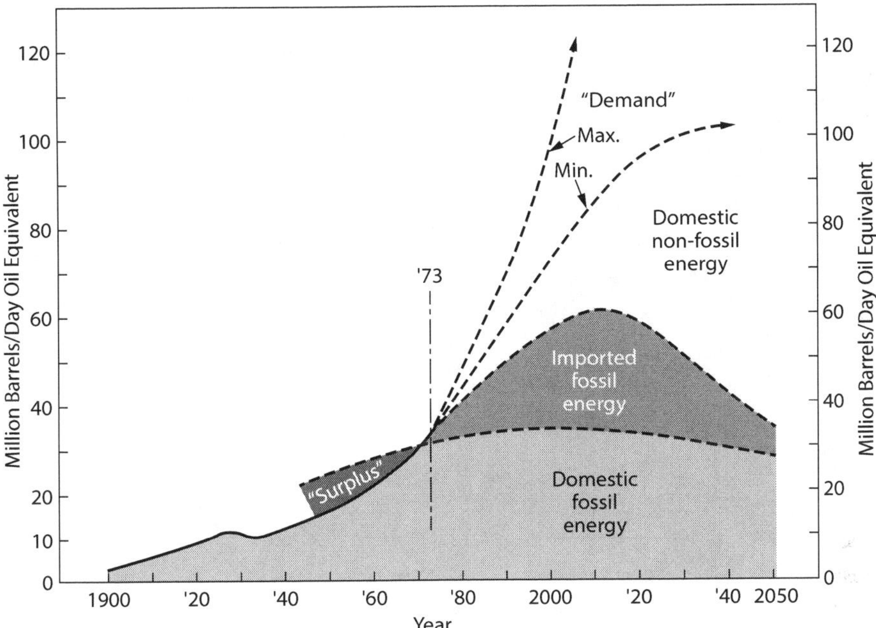

Figure 3. The Joint Committee on Atomic Energy rationalized deepening resource exploitation in the present with the understanding that US technical expertise would soon usher in a post-carbon era of nuclear power. Source: Joint Committee on Atomic Energy, *Understanding the "National Energy Dilemma,"* Fold Out P. Stanford University has made documents relevant to the JCAE available to the public at https://exhibits.stanford.edu/atomic-energy.

underwritten such progress, and their slow decline posed humanity's greatest challenge: to manage the transition to a different energy regime while maintaining the fruits of modernity. The newly minted historical period of the Fossil Fuel Age—bounded by the rise of coal during the Industrial Revolution around 1800 and the fall of petroleum in the 1973 oil embargo—seemed to be sputtering as oil approached exhaustion and diminishing returns.[78] One *Washington Post* report described a "horrified audience on Capitol Hill" faced with "ghastly" options for a post-carbon future as they realized the nation formerly awash in cheap oil would be utterly transformed, from its style of life to its role as a world power.[79] Others treated the end of the Fossil Fuel Age as a threat to the American commitment to social progress, to the New Deal, and to the suburban dream.[80] A few months of oil scarcity accompanied by years of commentary drastically altered perceptions of the past and future of US social and economic development.

Assessments of the energy crisis as a historical marker were often quite melancholy about the implications of leaving behind the Fossil Fuel Age. Many feared that the postwar period was a singular period of human flourishing that would be difficult to recapture after the transition that seemed to be underway. Economist Emile Benoit praised this new past as an "Age of Plenty" in which millions of people lived above mere subsistence for the first time in human history and looked askance at a future without cheap oil as a "coming age of shortages."[81] In his National Book Award–nominated *Awakening from the American Dream* (1976), former civil servant Rufus Miles Jr. called energy the "taproot of affluence" and described the crisis as the closing of a second frontier.[82] A study published by the Oak Ridge National Laboratory in 1982 similarly warned that "human progress and human cultural development" require large amounts of energy, though it agreed that the current fossil fuel systems had to give way to a cleaner and more secure system.[83]

Elegies for the passing of an age of easy abundance extended to the postwar suburban dream and the progressive impulses behind the New Deal and the Great Society. By the end of the 1970s, it was clearly understood that suburbia relied on the steady and inexpensive flow of gasoline and that rising energy prices and uncertain supplies could spell its end, not that this perception did much to stem the rising tide of suburbanization across the country. This was how one writer interpreted the 1979 trucker strike in Levittown, Pennsylvania, which ended in rioting and violence: "Like trapped animals pushed against the wall, the gas-guzzling dreamers of America's Levittowns have little left but frustration."[84] The energy crisis had demonstrated the existence of limits on presumed middle-class freedom of mobility, which frustrated the white working- and suburban middle-classes whose wealth had been steadily eroded by inflation, and undermined the hope of social mobility rooted in cheap energy.[85] Some commentators divined in this challenge to white suburban affluence a further challenge to the assumptions of the New Deal, after the initial assault by Nixon's Silent Majority. Pontificating on Ronald Reagan's election in 1980, historian and Kennedy liberal Arthur Schlesinger Jr. argued: "The commanding domestic issues of our time—chronic inflation and the passing of the age of low-cost energy—are novel issues and demand novel remedies." But after nearly a decade of hapless attempts to formulate effective energy policies, progressive faith in the efficacy of federal intervention to promote the common good seemed to be a mirage. Thus, Schlesinger highlighted the emergence of "a 'neo-liberal' school" of thought that opposed government intervention that could fundamentally reshape US political life.[86] Economists like

Paul W. MacAvoy and Morris Adelman tended to agree, often using the energy crisis to critique Keynesian intervention in favor of market deregulation.[87] Indeed, many pro-market economists hoped that the crisis signaled an end to the New Deal, Keynesian liberalism, and all they stood for.

It is important to note that melancholy energy crisis narratives about the end of a fossil-fueled Golden Age developed within the politics of race in the United States. Most media coverage focused on white middle- and working-class concerns, particularly those of postwar suburbia, which was largely a white world enmeshed with the power and prerogatives of the Cold War security state. Energy scarcity threatened the future of white suburbia at the same time that the civil rights and Black Power movements, the American Indian movement, and others were questioning the social and political presumptions of white America. It also undermined the ability of the United States to engage in the Cold War to defend the American "way of life." The ascendency of Nixon and the Silent Majority had articulated some of these fears in 1968 and 1972, and now the energy crisis appeared to deepen the threat by undermining the material conditions for US hegemony abroad and white hegemony at home.

White anxieties thus dominated energy crisis discourse, particularly the latter's nostalgia for the alleged stabilities of postwar life that were enjoyed mostly by white America. As soon as oil and gas prices skyrocketed during the first and second oil shocks, pre-crisis fossil fuels were reimagined as "cheap," abundant, and reliable. Media reports contextualized the crisis historically by noting that "cheap and plentiful energy" had "long been a driving force of the American economy. It used to be so cheap and so abundant that it was largely taken for granted and consumed with little thought for tomorrow."[88] Another report noted that US energies were "once considered an infinite supply."[89] Such accounts of the past were only possible with the benefit of hindsight and the injunction of crisis to understand the past anew, masking a more complicated and uncertain postwar history. They offered a simplified narrative of oil stability built on nostalgia for its carbon-based pleasures.

As postwar resource studies suggest, the US government worried about the longevity of its energy supplies as it embarked on the Cold War and its quest to spread capitalist democracy around the world. The authors of those studies did not think of energy as "infinite" but rather as a strategically vital set of resources to be counted, monitored, and preserved. Moreover, the relatively low oil prices that many Americans enjoyed before the 1970s resulted from policies and political decisions by federal, state, and local governments who decided that economic growth required cheap energy. They wrote laws encouraging oil consumption and production, stabilized historically volatile oil prices from falling too low, and kept oil prices from moving too high.[90] Yet the benefits of cheap energy were unevenly distributed by class and race. The effect of the rhetoric of past abundance was to make the comparatively cheaper energy prices of the past seem like an outcome of natural abundance rather than the product of conscious political decisions that were, nevertheless, accompanied by more uncertainty than nostalgic seventies narratives let on.

Recollections of a past awash in fossil fuels precipitated nostalgia for the material pleasures of this newly inaugurated Golden Age and laments for its passing. Many accounts decried the prospective extinction of the social rituals and visceral pleasures of car culture. High gas prices reduced middle-class spending on leisure, which had been a key social and psychic compensation for laboring under postwar capitalism.[91] As one Californian put it in 1979, "[I] can't afford to cruise as long as I used to: I only do it about an hour a night." Indeed, the *New York Times* reported, car enthusiasts had been forced to adjust "their passions to the reality of higher gasoline prices; some are beginning to acknowledge that time is running out on their way of life, yet they are still clinging to as much of it as they can afford."[92] California's car culture, the paper concluded, would never be the same in a world of high gasoline prices. During the second oil shock, the *LA Times* similarly observed: "Ask a Southern Californian what has changed for him during the last year, and there's a good chance he'll tell you what he doesn't do— go away for weekends; go to movies or restaurants as often; drive his car needlessly; pursue his hobbies."[93] With remarkable rapidity, then, the energy crisis transformed the postwar period into an economic Golden Age and a Fossil Fuel Age, both of which were passing away, but not without significant nostalgia for the pleasures and stability that they offered to those with access to them now beset with a pervasive sense of decline.

If the Fossil Fuel Age was ending, then a challenging period of transition was underway. Faced with the energy crisis, Shell petroleum geologist M. King Hubbert, one of the earliest and most prominent proponents of this new periodization, abandoned his initial optimism that nuclear energy could take the place of declining oil reserves and insisted on a low-growth future.[94] Hubbert supported this view by dividing world history into three periods. The first period stretched from the appearance of *Homo sapiens* to the Industrial Revolution, and its main features included low population growth, miniscule energy consumption, and a slow rate of sociopolitical change. The second period—now coming to be known

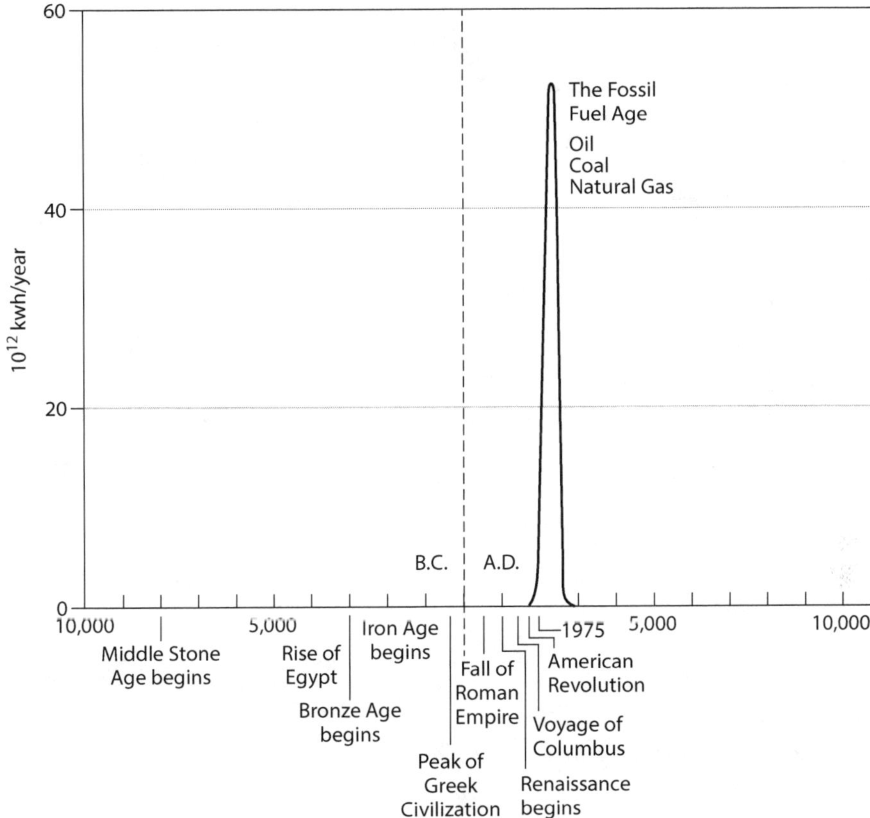

Figure 4. M. King Hubbert often depicted the Fossil Fuel Age as a historical exception. Source: From Office of Energy Research and Planning, Governor's Office, State of Oregon, Transition report, 1975, as appeared in Hayden, *The American Future*, 86.

as a Fossil Fuel Age—broke with the past and consisted of "continuous and spectacular exponential growth."[95] Hubbert argued that this phase depended on cheaply abundant fossil fuels, which no longer seemed likely. He therefore advocated a return to slow growth in a third, future era. Since energy consumption and population levels could not so quickly be halted, Hubbert understood the challenge of the future to be managing this demographic and economic transition to a steady-state society (fig. 4).

Experts seeking to make sense of the energy crisis rethought historical time in light of the crisis, which turned the American present into a time of transition. A transition was either underway or should have been, though most neoliberals preferred to talk about a transition to market-based policies rather than a planned transition away from carbon.[96] Certainty about an ongoing structural transition led to a flurry of debate about how best to navigate it and about the kind of society that might emerge on the other side. These debates sought to mobilize the crisis to support their social visions by articulating speculative accounts of desirable energy futures and the possibilities that they might open up to reshape US society.

Choosing the Future

Construing the present as the last gasp of a Fossil Fuel Age created space to imagine the society that should be the end goal of transition. Market advocates sought to revive US petroculture through economic deregulation, while environmentalists aimed to surpass it via a planned transition to alternative energies. Both visions threatened Keynesian liberalism. The variety of views that emerged demonstrate the flexibility of "crisis" to support competing narratives, which in the case of the energy crisis divided over different views about the economy of nature and the question of dependence. Would the future be a solar utopia, a carbon dystopia, or something else? Would it be a polluted future of oil abundance or one of frictionless nuclear abundance? And who would lead the nation into the future: the government, the people, or the markets? Whatever the answer, the energy crisis presented a fork in the road, "a time to choose."[97]

The decade's tidal wave of speculation about America's future energy system and its social and political implications included complex debates about the earth's carrying capacity, the desirability and nature of technological development and population and economic growth, as well as the values that different energy systems might serve. These debates revealed the fundamental futurity of the energy crisis as a crisis. Claims of crisis were grounded in demand continuing to climb and supplies failing to keep pace over time as much as present scarcities or economic effects. Present shortages always portended worsening scarcities that must be preempted by pursuing alternative futures.[98] As energy expert Daniel Yergin warned in 1978: "By 1985 or 1986 or 1987, if present trends continue, we'll be staring at an energy crisis far worse than the one we went through in the early 70s. And then our present boredom with the energy problem, and with the Carter Administration's efforts to cope with it, will seem like complacent sleep."[99] The energy crisis, in combination with stubborn inflation and a sense of waning American power, had made it possible to imagine regress, rather than progress, as the difference between the present and future. It echoed fears of nuclear or ecological apocalypse that raised the stakes of success and failure.

Despite this complexity, reformist energy futures discourse essentially posited two futures, as embodied by *Global 2000* and *The Resourceful Earth*: an environmentalist narrative about a low-energy future figured as humane and fulfilling or a market narrative of a high-energy future of postindustrial leisure based on free markets and an abundant economy of nature. Renewable energy advocates, back-to-the-landers, and others imagined a return to decentralized community life, envisioned either as a Jeffersonian republic or as neo-feudalism. Free market advocates argued that the market had produced past successes and would do so again in the future, if only the nation would return to it like a prodigal son.

Environmentalist narratives envisioned a low-growth future requiring deep structural changes in US society. Such views built on the notion of a "No Growth," "Zero Growth," or steady-state society that had reemerged in the United States and Western Europe in the mid-1960s alongside neo-Malthusian concern about pollution and population growth. Environmentalists, ecologists, and sympathetic economists argued that human population growth threatened to exhaust the earth's carrying capacity and overwhelm its ability to absorb pollution.[100] The energy crisis seemed to manifest these concerns, leading to interest in a steady-state society of stable population levels and stable energy consumption. The crisis did the same for other environmental critics, such as Barry Commoner, who pointed to overconsumption and technocratic hubris as the source of the nation's energy ills.[101]

In the mid-seventies, physicist and energy expert Amory Lovins characterized the choice that the world faced between a future of growth or equilibrium as a choice between two energy paths: a "hard path" and a "soft path."[102] Proponents of the hard path sought to expand available energy through coal, oil, and nuclear power, rather than "softer" alternatives like solar and wind. The assumption of the former was that social progress depended on high energy consumption, a view that was mainstream by mid-century.[103] Such high-technology solutions required forms of centralization and bureaucratic expansion that low-growth advocates abhorred.[104] In contrast, the soft path involved the creation of a distributed, low-tech, and more environmentally benign energy system. According to Lovins, it valued renewability, diversity, and flexibility, as well as scales and magnitudes appropriate to end uses.[105] Lovins and other low energy advocates agreed that this soft path future would require a decentralized network of locally controlled solar and wind power that would be socially revolutionary.

Most advocates of a low-energy future believed that their vision promised a more democratic, humane, sustainable, and equitable society than the high-energy society that the world was leaving behind.[106] Soft path technologies were said to distribute risk more equitably than nuclear power since millions of individual solar panels and wind farms would be more difficult to disable and much less likely to fail en masse than a single nuclear power station.[107] A decentralized republic—often couched in utopian Jeffersonian rhetoric—was also said to entail greater "consumer choice," freedom, and diversity where the hard path required centrally managed, expensive, large-scale, high-technology projects.[108] The latter were anonymous, uncontrollable, and politically inaccessible to the poor while the former were accessible to everyone, more equally distributing the costs and benefits of the energy system. The soft path seemed to be inherently more participatory and less coercive.

As Lovins argued, "Hard technologies are oriented toward abstract economic services for remote and anonymous consumers, and therefore can neither command personal involvement by people in the community they serve. Soft technologies, on the other hand, use familiar, equitably distributed natural energies to meet perceived human needs directly and comprehensible, and are thus, in Illich's sense, 'convivial' to choose, build, and use."[109] Here Lovins references the heterodox social critic Ivan Illich, whose critique of technocracy informed his analysis of hyper-industrialized energy and social structures. In *Tools for Conviviality*, published months before the OAPEC oil embargo, Illich had envisioned his ideal social future: a "convivial society" that used technology to support "human autonomy, freedom, and flourishing."[110] He, and others like E. F. Schumacher, echoed the 1960s critique of technology as alienating by arguing that industrial technologies dictated rather than served human life and so had to be abandoned. They recast modernization and progress as the pursuit of fewer but more-precise energies while relying on embodied energy to heal society and transform the self through dependence on nature and other people.[111] Blending a vision of decentralized democracy of greater consumer choice and social equality without centralized authority, low-growth futures confounded the progressive/conservative binary that is often used to explain the political changes of the 1970s.

These and other low-growth advocates used the energy crisis to persuade the public that their analysis of the relations between energy, society, and desirable futures offered the best way forward for the industrial West. They argued that the crisis exposed the impossibility of pursuing growth and social equality simultaneously, in part because its effects disproportionately hurt poor Americans and the world's less developed nations.[112] They insisted that higher energy consumption degraded social relations and social spaces by concentrating political power, corrupting political institutions, reducing choice, and enslaving individuals to energy dependence.[113] Illich, for instance, deconstructed high-energy transportation systems to expose how they bred dependence and disproportionately wasted the time and autonomy of the middle and lower classes.[114] He showed how energy regimes structured "the configuration of social space and life" by shaping how people spent their time, built their communities, and formed their identities as nodes in an accelerating transportation network.[115] What, then, might a more democratic, fulfilling, and equitable low-energy future look like?

Few low energy advocates painted detailed pictures of a soft-tech future, but they often discussed the qualities that such societies should have.[116] Most agreed that the low-energy future would slow down the pace of life and change but would also be more communal, equal, and free, in part because people would have to live closer to nature and to each other in order to thrive. Fulfillment would emerge in social relations and self-sufficiency rather than material accumulation or mobility. Scholar Gary Coates envisioned "a nonviolent, frugal, egalitarian society based on the renewable energy sources of sun, wind, moving water, and biomass."[117] Environmentalist Denis Hayes argued that a low-energy solar future could "dramatically affect the international distribution of wealth" by encouraging investment in poorer countries. Materialism and "frivolous consumption" would give way to new social values.[118] Counterculture icon Tom Hayden foresaw a "community of self-sufficiency" marked by peace and sustainability. This sense of a soft transition to a peaceful future was even implied in the gently converging lines of Hayden's graphic illustration (fig. 5), though it also recalled the strains of frontier individualism evident in Project Independence.[119]

The establishment of these new values, structures, and environments would generate new forms of economic life. Physicist Fritjof Capra, ecologist H. T. Odum—who anticipated the abrupt arrival of the steady-state society—and others, predicted that energy would replace money because the energy cost of any given activity mattered more than money.[120] Odum's "energy theory of value" elevated the principles of systems ecology and energy flows over neoclassical economic principles of money flow and monetary policy as the primary prism for managing the economy.[121] He saw the shift to this careful energy accounting as imperative in light of the continued degradation of its fossil energy sources. A steady-state future would be foisted onto the United States, which would have to adopt less energy-intensive agriculture, commerce, education, and values.[122]

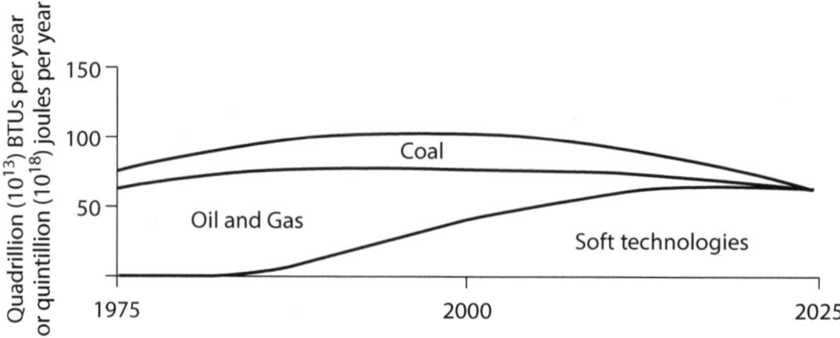

Figure 5. Gentle lines illustrate a transition to a low-growth future. Source: From Anna Gyorgy and Friends, *No Nukes*, as appeared in Hayden, *The American Future*, 111.

Not every vision of the low-energy future was so hopeful. Some foresaw a sudden crash driven by demography, famine, nuclear war, or peak oil and argued that a low-energy future entailed the end of democracy and the potential for a new dark age.[123] The political scientist and environmentalist William Ophuls envisioned a return to feudalism: in an era of "ecological scarcity," liberalism would give way to communal values of frugality and duty, as well as an aristocratic authority to legislate energy temperance.[124] In this, he echoed the managerial and coercive prescriptions of neo-Malthusians like Paul Ehrlich and Garrett Hardin. The geologist Earl Cook, drawing extensively on the work of Fred Cottrell, agreed that democracy was unlikely in a low-energy society, in part because it did not accommodate the time horizons needed to manage the resources for long-term sustainability. Cook seemed to be satisfied with this, however, because he saw in "high-energy society" the source of many 1970s social ills, including obesity, crime, and drug culture.[125]

These imaginaries of a low-energy future drew on an emerging conception of energy as a property of systems and on the decade's tendency to fetishize energy as a fundamental factor for all of life. Both currents flowed out of postwar systems ecology. H. T. Odum considered energy to be so foundational to the nature and dynamics of all systems that he declared it to be "the source and control of all things, all value, and all the actions of human beings and nature."[126] Writing in *Scientific American*, physicist Freeman J. Dyson similarly wrote that nature of energy "lies at the heart of the mystery of our existence as animate beings in an inanimate universe."[127] It is not hard to see Janet Roitman's notion of crisis as a form of secular meaning making in the energy crisis when energy, rhetorically at least, attained such mystical status as a unitary explanation for many different phenomena. Indeed, policy makers and experts began to talk about energy as the key "to knowingly 'invent'" the future.[128] A Rockefeller Foundation report expressed the emerging view of energy's importance: "Most of the world is going through a sea-change. The ultimate effects of the current energy transition will represent changes in kind rather than degree; not simply a shift from one energy to another, but a change in thinking about national and social policy, in value systems, in the world balance of power, in how we perceive the future."[129] But if energy was so fundamental to social change, the key to a different future, then low-energy gurus considered it first to be a property of systems, and second to be an embodied force. The system seemed to be the key site of energy transition. By designing different systems, societies could encourage sustainable uses of energy.

These environmentalist prescriptions assumed a limited economy of nature. Low growth was essential if humanity was to avoid overwhelming natural systems. This view drew on the sense of limits in postwar neo-Malthusian analyses of population and resources, though its intellectual foundations also included the systems ecology and ecological economics of the 1960s and 1970s, which insisted that human systems, such as national economies, cities, and households, had to be redesigned to accord with the laws of nature that governed all systems, and that set limits on their growth and development.[130] Unlike mainstream classical and neoclassical economics, ecological economics prioritized the laws of nature, particularly in physics and biology, in its account of how human economies function. More particularly, it held that human economies, and therefore the discipline of economics, can *only* be understood as subsets of the earth's ecological systems and their laws. Economics had to subsume ecology.[131]

Low-energy imaginaries of scaled-down, decentralized energy systems marked by self-reliance and authentic community inspired thousands of people to reorganize their lives to use less. As David Shi notes, the pursuit of the "simple life" has periodically dotted the cultural landscape of the American past, but its 1970s manifestation tended to adopt "a more global perspective and an almost apocalyptic immediacy."[132] Adherents perceived the energy crisis as a spur to adopt lifestyles of "self-restraint and self-reliance."[133] Inspired by writers like Wendell Berry, the 1970s simple lifestyle movement embraced low-energy futures as a way to resist the culture of growth and revive a Jeffersonian vision of America. In his call to return to a life of meaningful labor on the land, Berry noted that energy is "not just fuel" but is rather "a powerful social and cultural influence." The nation's energy problem, he argued, began when agriculture industrialized, rendering farmers dependent on nonrenewable fossil fuels. This process hollowed out rural communities and exacerbated the abuse of the land.[134] In order to spread these ideas and equip people to act on them, dozens of periodicals of advice, inspiration, and practical tools appeared in the 1970s, including the *Whole Earth Catalog, Rain, Organic Gardening,* and *Mother Earth News.*[135]

Mother Earth News is a particularly compelling example because it reveals the deep anxiety about dependence that marked popular environmentalist lifestyle literature in the 1970s, which often advocated a species of libertarian future. Established in 1970 by the husband-and-wife team of John and Jane Shuttleworth, *Mother Earth News* offered practical advice on ecological, back-to-the-land living alongside articles and interviews that articulated the heterodox intellectual foundations for the movement. The magazine's "Plowboy Interview" series exposed readers to the ideas of environmental thinkers like H. T. Odum, Wendell Berry, Paul Ehrlich, Amory Lovins, Edward Abbey, and a host of others advocating a retreat from industrialism and growth. Its many how-to articles explained strategies and tools for self-reliance in food and energy production, including homesteading, livestock care, small-scale solar energy production, solar home design, and economic self-sufficiency.

The editors and readers of *Mother Earth News* lamented what they took to be the cultural decadence of modern America, calling the energy crisis "a *gluttony* crisis . . . [and] an *addiction* crisis."[136] Their solution was to live self-reliantly and independently on the land—a view that contained within it deeply felt disdain for reliance on large, impersonal entities like big oil and the federal government, as well as disdain for the reliant themselves. As one article on energy self-sufficiency tellingly concluded: "With sun-heated water flowing through baseboard heaters to provide general home and cellar heat, and with wood stoves to really warm up rooms in use, we will be able to forget Exxon, Mobil, and the Arab oil sheiks for good."[137] Clearly, the vision of egalitarianism and peace of the environmental narrative coexisted with a libertarian strain of disdain for dependence rooted in larger national anxieties about oil dependence and the many other crises besetting the United States.[138] Within that atomized, resentful vision, as Sam Binkley has argued, lay the seeds of the neoliberal ascendency of the 1980s, as disdain for dependence fostered disdain for government and, eventually, enthusiasm for the potential abundance and alleged freedom of turning to the market as source of self- and social determination.[139]

While environmentalists advocated varying visions of a low-growth energy future, competing narratives emerged of an abundant energy future tied to the promises of the free market rather than planned decentralization and degrowth. Such narratives echoed the interests of the federal government and corporations, who spent millions of dollars to restore energy abundance through coal gasification technology, synthetic fuel, oil and gas exploration, pipeline construction, hydrogen power, and nuclear power. Experts justified these pursuits by arguing that only a secure, abundant, and inexpensive supply of energy could power the economic growth necessary for the survival of capitalism, democracy, and the realization of human potential. Many pro-growth advocates favored a free market approach, announcing their faith in the ability of the market to eliminate scarcity with anticipatory rhetoric about the possibility, abundance, and adaptive potential that markets offered.[140] The abundance of nature was flexible and open to humanity, if only governments and other bureaucracies would get out of the way.

The neoliberal economists and businessmen who imagined energy-abundant futures conceived of the economy of nature as unlimited—a position that shared the optimism of some scientists and futurists from the period.[141] The former aimed to integrate humanity into the inherent abundance of nature rather than conforming to a notion of nature's limits. For instance, the popular microbiologist René Dubos, praised by the *New York Times* as an optimistic salve to American "ears beleaguered by negation and despair," preached the power of human creativity to develop technological and cultural solutions to the energy and environmental crises by working creatively with nature. Dubos decried doomsayers and questioned the assumption that more energy consumption inevitably improves human welfare.[142] He insisted that the energy crisis would be "a blessing" if it compelled Americans to "develop ways of life that encourage fuller expression of the adaptive and creative potentialities that are present in us and in nature."[143] By emphasizing adaptive potential as an inherent evolutionary trait, Dubos interpreted the energy crisis as a chance to experiment, to adapt, and to grow.

Dubos's pro-growth rhetoric, although intended to praise qualitative over quantitative growth, dovetailed with neoliberal notions of the market's ability to ensure abundance.[144] For instance, he claimed that the "limits to growth" model offered a false ontology of natural resources, which in reality are relative to the time and socioeconomic system in which they are created and used. Since the evolution of human societies shapes human activity, different eras required different kinds of resources. "Each age," he said, "creates the resources that it needs." So if oil runs out, human ingenuity would eventually create another kind of resource to be a prime mover. Awareness of an energy crisis would generate solutions to it. Nature would always afford a new resource to human ingenuity because both were pregnant with creative potential. Growth, understood in Dubos's terms, had "no discernible limits."[145]

Emerging neoliberal narratives of resources and markets shared Dubos's optimistic focus on creative adaptation and the abundance of nature. By the end of the 1970s, economist Julian Simon and others were arguing that "finiteness" was "not a problem" in energy as long as markets remained free. Simon specifically claimed that the human imagination was the world's "ultimate resource" and that it would solve the energy and any other kind of crisis by tapping the wealth of nature that Dubos so fondly described.[146]

To offer support for his optimism, Simon referenced the controversial astrophysicist Thomas Gold, who proposed in the late 1970s and 1980s an abiogenic theory of petroleum as an unlimited resource that is constantly created deep in the earth's crust.[147] Gold concluded that "if deep gas is plentiful and can be discovered in many parts of the world and in many geological settings, the economic outlook would be greatly improved and many countries would breathe more freely when they are relieved from the obligation to build or expand a nuclear power industry in a hurry, or from the need to depend on imported fuels."[148] Gold called this his "deep earth gas hypothesis," in which earthquakes release gas from the earth's mantle, including methane of nonbiological origin, that could become an important source of fuel for the global economy.[149] Maintaining the optimism that underwrote neoliberal energy panaceas, Gold speculated that humanity might discover "virtually inexhaustible reservoirs" constantly being replenished. By the time humans could exhaust natural gas supplies, he said, "one may expect that totally different and commercially viable sources of energy will have been discovered" or new and better drilling techniques would be developed to make possible a "deeper horizon" of energy exploitation.[150] Such optimism about the abundance of nature to afford an endless roll call of natural resources for human purpose was fundamental to the neoliberal notion of the market, which may explain the public attention that Gold's controversial ideas received as energy crisis discourse shifted from scarcity to possibility.[151]

Gold's optimism about the economy of nature brings us back to the *Global 2000* and *Resourceful Earth* reports, which clashed over this question. The former narrated the energy crisis as a crisis of scarcity and imagined a low-energy, low-growth future rooted in a limited economy of nature. The latter joined think tanks like the American Enterprise Institute in praising the efficacy of markets to unlock nature's bounty. For both, the energy crisis confronted the United States with a choice about the kind of future that the nation, and the world, should pursue.

Competing narratives offered visions of a better future, with either a more abundant interior and communal life, or a life of endless energy to serve human needs. Neoliberals, like Milton Friedman, Herbert Stein, William Simon, and Morris Adelman, stridently resisted what they saw as the inherent and dangerous pessimism of the low-growth movement in favor of free market abundance.[152]

Market narratives of the future drew strength from the optimism of those who believed nature to be an unlimited source of value for unlimited growth. Herman Kahn's work throughout the 1970s boldly predicted that people would be "numerous, rich, and in control of the forces of nature" in two hundred years.[153] Population and economic growth would continue for a long time, eventually slowing down not due to limits but rather to declining demand. Energy, economic, and population issues were all "basically solvable or resolvable in the near and medium term future," not intractable issues that would only worsen over time as many no-growth scenarios predicted. Energy costs would continue to decline as they had historically, and the introduction of cheaper and more efficient energy would loosen the bond between energy consumption and economic growth. Indeed, energy abundance was the key to the security of this future of thirty billion people living in comfortably wealthy democratic countries, which would be achieved through solar, hydrogen, and geothermal energy. Kahn took technological progress to be both certain and far easier to obtain than the more formidable challenge of cultural adaptation to such wealth and comfort. He believed that creating such clear and optimistic images of the future was essential to achieving consensus and commitment to human progress. "We need," he declared, a future "that opens possibilities rather than forecloses them."[154] Later in *The Resourceful Earth*, Kahn echoed Dubos by arguing that material progress was limited only by the human perception of nature as limited and by the state's needless reinforcement of those perceptions. The energy to fuel progress awaited liberation by technology, if only the state would stay out of the way.

The energy crisis inspired US leaders and experts throughout the 1970s and 1980s to worry about the crumbling foundations of American prosperity, but this process eventually reoriented US political culture toward speculative faith in the malleability of the earth and the efficacy of the free market to unlock its value. Shortages, as economist Philip Gramm had declared during the 1973–74 oil embargo, could not exist in free markets.[155] He, like many free market advocates in the 1970s, invoked the nineteenth-century whale oil crisis as a historical precedent to frame the 1970s energy crisis as just another textbook case of market dynamics. As Gramm insisted, Americans owed "the benefits and comforts of the present era to free enterprise and the scarcity of whales."[156] Popular Wall Street financial analyst Raymond DeVoe Jr. turned this insight into a virtual law of history, arguing in *Barron's* that the energy crises of the past, from scarcities of Roman slaves to whale oil, offered opportunities for technological innovation and progress.[157] In scarcity, therefore, lay opportunity: potential waiting to be unleashed by human creativity. By the end of the 1970s, the New York Mercantile Exchange was well on its way to establishing an oil futures market, which created economic value out of speculation about future oil prices in a volatile world (see chapter 5). The location of value in this imaginary was neither the traditional nor the bio-economy—it was in speculation and risk management, girded by optimistic faith in the possibilities of the market and nature.[158]

The energy policy of the Reagan administration encapsulated this risky optimism. Rooted in a view of the energy crisis as a manifestation of government meddling, Reagan's policy deregulated the energy sector, claiming that "there is good reason to be optimistic about the future effects of this policy on the discovery and production of domestic oil reserves" because freer markets would stimulate exploration and innovation while eliminating scarcity in the long term. Yet the administration also acknowledged its own uncertainty about the future: "The one thing that is certain about the future is the exact path of energy developments and markets is uncertain." The only way forward, therefore, was to establish free markets for energy. This entailed a risk that the administration was willing to take because of what the nation's oil resources might be—the hope that its liberal program of leasing on the Outer Continental Shelf, as well as oil shale and tar sands lands, would unlock the abundance of nature. Tied to this, however, was the anticipation of chaos and disruption, which shadowed the administration's focus on emergency preparedness and stockpiling for uncertain futures that debates about the meaning of the energy crisis had made palpable.

Conclusion

Crisis is not a self-evident reality to be discovered "out there" in the world. Crisis is, instead, an interpretation of events. As such, it is a deeply political and normative claim.[159] For Janet Roitman, the ubiquity of crisis is a symptom of contemporary suspicion of transcendent metanarratives: it enables those who distrust appeals to transcendence or reason to demarcate what is "historical" from the undifferentiated stream of past events.[160] In modernity, it is the claim of crisis that structures historical knowledge and grounds demands for change. But as Roitman also points out, the centrality of crisis to contemporary critique enables the concept itself to become "a latency" that escapes scrutiny.[161] We often rely on crisis to make claims about the world without reflecting critically on the category of crisis itself.

My concern in this chapter has been to investigate how one crisis functioned discursively as an object of knowledge to ground competing claims for change. I used the tripartite schema of crisis devised by theorist Reinhart Koselleck to clarify some of the temporal semantics by which crisis makes meaning out of time. In the first place, crisis discourses imagine that History is executing judgment for past wrongs, a clear secularization of Christian eschatology that compels the question of what went wrong in the past to bring about a crisis moment in the present. Secondly, crisis narratives reassess and re-periodize the past in light of the present.[162] Questions of progress are central to this frame, in which it can be unclear if crisis is the engine of progress, or if progress is the engine of crisis.[163] Finally, crisis posits a "final decision." It becomes a sign that History is on the cusp of a new, salvific time. The theological also haunts this mode, wherein crisis is the judgment that inaugurates a new and better historical phase. In all three cases, crisis claims create meanings that are always contested, as are the futures they posit. My aim in organizing the cacophony of energy crisis interpretations in this way is to highlight the centrality of temporality to energy crisis discourse. As much as the energy crisis concerned geopolitics and energy policy, it also opened space to reassess the past and the future of the American project at a crucial moment.

The competing market and environmentalist narratives of the energy crisis suggest a politically flexible crisis that could be mobilized in support of conflicting visions of the world. This chapter explored how these competing narratives differed over the nature of the crisis and of nature's economy but nevertheless shared Koselleck's interpretive categories. By articulating their understanding of the crisis, different institutions, experts, and individuals tried to make "crisis" do important moral and political work to address the challenges besetting the United States. In keeping with the temporal nature of crisis, their interpretations always either reimagined the past or articulated futures rooted in an anticipatory understanding of the energy crisis.

The environment/market binary used here also alludes to the main fault lines in the ascendency of neoliberalism in US political culture, which confounded the progressive/conservative binary that occupies the historical literature concerned with the decline of the New Deal and the rise of the New Right. The evolution of the energy crisis assemblage through these debates, where progressive critique and the market intersected, suggests a neoliberal realignment of US political culture that exceeded categories like "progressive" or "conservative." Both energy crisis narratives undermined Keynesian liberalism as a failed project. Environmentalists supported the use of state power but rejected growth as a goal and disavowed the postwar period as any kind of model for the future. Market advocates, on the other hand, embraced the goal of growth but rejected the Keynesian paradigm for achieving it. The next chapter turns from interpretations of the energy crisis toward another element of the assemblage: the energy conservation ethic as a widely proposed solution to the energy crisis, but again for competing ends.

"A Vibrant National Preoccupation"

The Energy Conservation Ethic and Market Forces

If supplies fail, try demand.   —*NEW YORK TIMES*, OCTOBER 1973

In 1979, Conservation Foundation president William K. Reilly declared that "people who may be able to agree on virtually nothing else . . . can reach consensus on the need to conserve energy." He thought that most Americans had discarded their earlier skepticism about the nature of the crisis and were now willing to act. The recurrence of oil shortages after the Iranian Revolution had created a new sense of urgency in the public that presented "an opportunity for the conservation community to return to its historical concern about resources, and to clarify the central role of resource conservation to the long-term health of society."[1] By uniting issues of oil dependence, inflation, and the environment with a popular desire to restore US power, Reilly thought, the energy crisis could mobilize a broad constituency for conservation. *Newsweek* agreed, declaring energy conservation to be a "vibrant national preoccupation."[2] Indeed, an individualized and consumerist notion of energy conservation emerged in the mid-1970s as a key way to respond to the energy crisis. Republican and Democratic administrations, environmentalists, and corporations alike reproached American consumers for what prominent bureaucrat William Simon called their "wastrel ways" and chided them to change by adopting an "energy conservation ethic."[3] This chapter explores why such a broad consensus developed around a consumer-oriented notion of energy conservation as a solution to the energy crisis and why it fell apart so quickly in the early 1980s.

There are two prongs to my argument about the 1970s energy conservation debate. First, a consensus developed around energy conservation because the notion of a conservation ethic was sufficiently broad to contain two distinct sets of ideas about its nature and purpose. Some of its supporters embraced what might be described as an ecological paradigm, and so they viewed the conservation ethic as a route to sustainability through permanently reduced energy use. Others followed an explicitly nationalist paradigm, advocating a conservation ethic as a means of eliminating dependence on foreign oil in order to advance the economic and geopolitical strength of the United States. This latter paradigm recalled early twentieth-century conservationism by stressing national vitality as the raison d'être for conserving resources.[4] Although its focus on consumption reflected the influence of environmentalism, its existence demonstrates the ongoing relevance of conservationism in the "environmental decade."[5] Together, these ecological and nationalist paradigms fostered broad support for an energy conservation ethic, bolstering later neoliberal emphases on austerity, but their competing visions of its purpose limited the durability of their coalition.

The fact that members of both the ecological and nationalist paradigms sought to reduce energy consumption and announced the need to value energy resources differently to achieve that end provides the foundation for the second line of argumentation—namely that the concern with individual behavior and deregulating energy prices that permeated the discourse of the 1970s conservation ethic reflected an ongoing neoliberal shift in US political culture. The energy conservation ethic, like the pervasive "energy crisis" discourse, conflated different forms and crises of energy, often imagining lighting (powered by coal, hydro, oil, or nuclear) and reduced driving (powered by oil) as responses to a single crisis. Dominant articulations of a conservation ethic by government officials, corporate representatives, and wealthy interest groups tended to focus attention on the reformation of individual habits to solve the crisis. After failing to enlist sufficient voluntary reductions in energy use, adherents from both sides came to support a market-based approach to conservation, most prominently in oil price decontrol. If, as Meg Jacobs has argued, the energy crisis empowered the deregulatory designs of Washington conservatives, the energy conservation ethic facilitated what Reilly called an "unholy alliance" between environmentalists and pro-market advocates in favor of free market pricing to discipline America's unruly energy consumption and to fund the further development of US energy resources.[6] For some, the energy conservation ethic was one of austerity, aided by environmentalist critiques of affluence, that promised abundance over the long term. When

Ronald Reagan decontrolled oil prices in 1981, he declared that market pricing would foster "prudent conservation and vigorous domestic production" to endow the United States with economic security and vitality by valuing its energy resources at their "true value."[7] As prices fell and supplies rose in the 1980s, the conservation nationalists seemed to have achieved their limited goals, and the consensus on energy conservation dissolved.

An Energy Conservation Ethic

Debates over exactly what energy conservation entailed ranged widely in the 1970s. Proposals included everything from radical new socioeconomic arrangements and alternative energy technologies to incentives for building insulation and more efficient production methods.[8] Underlying all of these proposals, however, was the notion of a new *ethic*. The problem of energy conservation, wrote one expert, could not be reduced to engineering better technology—it had to address "human attitudes and actions."[9] Whether labeled as an "energy ethic" or an "energy conservation ethic," this framing dominated public discussions about reforming energy use in the 1970s—particularly after the oil embargo of 1973 and 1974—and asserted the need for Americans to internalize new habits and values around energy consumption.[10] At its core, the energy conservation ethic posited that US consumers must become newly aware of their energy habits in order to return to the habitual frugality of earlier generations by monitoring and revising quotidian behaviors like driving, doing laundry, or using electric lighting. The challenge was to achieve frugality before the necessity that constrained earlier generations reared its head once again.

Concern with energy conservation has a relatively short history. As Cara Daggett points out, energy was "discovered" only in the 1840s.[11] Prior conservation efforts, from French forests to American landscapes, thus did not speak in terms of conserving "energy." In the mid-nineteenth-century United States, diplomat and conservationist George Perkins Marsh articulated the need to conserve the nation's forest resources but did not frame his concerns in abstract terms of *energy*. His influential book *Man and Nature* (1864) focused on productivist uses of nature—chopping down trees for fuel and construction, practicing agriculture—and warned that the cumulative effect of human resource use demanded prudent resource management for the sake of future generations.[12] *Energy* conservation entered the picture more explicitly in William Stanley Jevons's warning that coal exhaustion posed a long-term threat to the British Empire.[13] A few decades later, the principles of thermodynamics and the science of work extended energy conservation concern to managing the physiological energies of laboring human bodies to achieve maximal productive efficiency.[14]

The conservation movement of the early twentieth century built on these precedents in its efforts to reduce waste and to manage the stocks of US energy resources for their "fullest" use and longevity. For Gifford Pinchot—the first head of the US Forest Service, adviser to President Theodore Roosevelt, and later the governor of Pennsylvania—the purpose of conservation was to ensure resource abundance for the sake of national vitality and future generations.[15] It entailed an ethic, to be sure, but one largely aimed at productive practices beyond the everyday purview of most citizens. The same was true of the regulatory system governed by the Texas Railroad Commission from the 1930s to the 1960s, which conserved US oil resources by way of production quotas meant to stabilize oil prices.[16] The goal of all of these early energy conservation efforts was to manage production to prolong resource availability for the sake of national strength.

The notion of an energy conservation ethic that emerged in the 1970s was different; it was less exclusively productivist and tied more directly to everyday experience. It resembled the thinking of the forester and proto-environmentalist Aldo Leopold, who held that "the practice of conservation must spring from a conviction of what is ethically and esthetically right." Leopold most comprehensively outlined his approach in his postwar classic of environmental writing, *A Sand County Almanac*, which described what he called a "land ethic" focused on reforming values and behaviors for ecological ends. Speaking in the context of land management, in which conservation meant "a state of harmony between men and land," Leopold wanted individuals to integrate limits to freedom into their values, desires, and motives. He saw modernity as an alienating force that created disconnection and ignorance about the land and fostered its abuse. Only by restoring "intense consciousness of land" would individual citizens be able to learn to love it anew and to internalize the principles of a land ethic that would help them to choose sacrificial acts for the sake of the collective.[17] For Leopold, a conservationism rooted solely in economic self-interest was insufficient. People had to learn to care for the land's resources and to act out of that sense of personal responsibility. The agrarian writer Wendell Berry, who rose to prominence when the Sierra Club published his book *The Unsettling of America*, took up Leopold's mantle in the 1970s. His work offered a holistic vision of the integration of culture, agriculture, and human communities, arguing for agriculture as a dignified cultural activity that must prioritize the integrity of the land in its decisions about how to use soil, machinery, and energy.[18]

During the energy crisis, intense consciousness of energy consumption joined Leopold's "intense consciousness of land" as a driving force in US environmental politics. In light of the power shortages, gasoline lines, and oil price shocks that informed energy crisis discourse, consuming less became both more urgent and more practical. Activists, experts, and government officials seized the sudden realization that modern life relied on enormous amounts of energy to reform the once-valorized consumer of postwar America into a frugal energy conserver. Beneath the superficial similarity of their calls to rethink energy consumption, however, lay radically different aims. For some, expanding Americans' consciousness of the energy flows supporting everyday life facilitated a broader effort to live according to ecological principles, in the spirit of Aldo Leopold. For others, restraining energy consumption meant disciplining everyday behavior as a form of patriotic sacrifice, shifting Pinchot's focus on production to consumer behavior. Although these aims were not inevitably at odds, their divergent ideological foundations pointed them toward very different ends.

The Ecological Paradigm for Energy Conservation

The ecological paradigm contained a diverse array of figures and institutions that nevertheless cohered around the necessity of social and political change to create a conservational society of permanently reduced energy consumption, an antecedent to today's notion of sustainability. Adherents to this way of thinking included ecological economists, systems theorists, ecologists, physicists, and philosophers, as well as a number of citizen and community groups. Although they approached the problem of energy consumption from different points of view, they all criticized the primacy of economic growth in the West and the fossil-fueled consumption that supported it and its attendant ideas, values, and cultural practices.[19] Their call for transformation drew strength from growing environmental movement of the 1970s and a shared interpretation of the energy and environmental crises as symptoms of the fragility and unsustainability of industrial capitalism. The system had to be changed, and the energy crisis provided an opportunity to pursue fundamentally new directions. As a sign of a deeper ecological crisis of overconsumption, they thought, the energy crisis could be used to motivate Americans and their government to consume less.

Adherents of the ecological paradigm posited that adequate and lasting energy conservation required a new social order that valued energy differently, abandoned growth, and required fewer of the earth's resources.[20] By this they often meant the creation of some form of steady-state society—a society that aims to maintain constant levels of population, capital, and consumption, rather than constant growth. The concept has roots in classical economics, but its twentieth-century form prioritized the management of energy in the context of the emerging discipline of ecological economics. As described by Herman Daly, a leading ecological economist at Louisiana State University, a steady-state society aims to balance inflows and outflows of "matter-energy." It renounces the massive fossil fuel inputs that transformed twentieth-century life and works within the laws of thermodynamics and ecology.[21] Those in favor of a steady-state future imagined future growth and development in qualitative, rather than quantitative, terms: of knowledge, happiness, and community. Many proposed shifting to decentralized economies that produced durable products through the efficient use of resources to preempt the energy transition that the energy crisis portended.[22] But "all this," wrote the Romanian American economist Nicholas Georgescu-Roegen, "boils down to the need for a change in our values."[23] A steady-state society could not merely be a technocratic innovation; it had to come from new ways of thinking and feeling about the place of human economy in nature's economy.

This appeal to values often implicated everyone, much like US environmentalism more broadly in the 1970s. The famous Earth Day poster featuring the popular *Pogo* comic strip declared, "We have met the enemy and he is us," universalizing the ethical burden of environmental responsibility, shifting from the conservationist's focus on production to the environmentalist's concern with humanity's impact on the environment.[24] Proponents of the ecological paradigm framed the relationship between energy and the environment similarly: *everyone* bore some responsibility for the energy crisis and all were implicated in the reimagined system of values expressed in the energy conservation ethic. In contrast to the technocratic and anthropocentric ethics of growth-oriented societies, a steady state called for an eco-centric ethic in which homeostasis, holism, limits, and sustainability are ends in themselves throughout society.[25] Or, as Georgescu-Roegen suggested, "Love thy neighbor as thyself," would be exchanged for an eco-centric alternative: "Thou shalt love thy species as thy self."[26] Conserving energy seemed to embody and serve all of these ends. It was a universal moral concern.

The moralizing and universalizing sense of the energy conservation ethic dovetailed with 1970s critiques of US consumerism advanced by conservative intellectuals, environmentalists, New Left activists, and others. For instance, in 1970, students at San Jose State College buried a brand new Ford Maverick to protest the impact of overconsumption on the environment and to call for new ways of living with the earth.[27] African American students criticized the stunt as a privileged waste of money, and African American community leaders called for a broadening of the environmental movement to consider urban issues.[28] Nevertheless, the car burial stunt, and many aspects of Earth Day protests on April 22, 1970, targeted US consumerism as a social, political, and moral sin to be purged. Several prominent intellectuals, including Daniel Bell and Christopher Lasch, also developed a critique of consumption as inimical to the moral, cultural, and political health of the nation. They argued that overconsumption threatened the habits of delayed gratification essential to capitalism and healthy community life.[29] Jimmy Carter distilled their views and broadcast them to a larger audience in his July 1979 "crisis of confidence" speech, in which the president scolded Americans for prioritizing "owning and consuming things" over community and self-sacrifice. Similarly linking energy and consumerism, an enduring and widely read essay by Ivan Illich argued that a conserving society would entail "liberation from affluence" and from the "solitude of plenty." He saw the low-carbon reorientation of life as key to rectifying the alienation, inequality, and dependency inherent in fossil-fueled societies.[30]

Key figures in the ecological paradigm argued that the thrifty habits and values of the conservation ethic need to be paired with "soft energy path" technologies to make energy transition feasible.[31] As we saw in chapter 2, Amory Lovins introduced the term "soft energy path" in 1976 to describe energy technologies based on efficiency, renewability, and appropriate scale over against centralized, large-scale, and new-renewable "hard energy path" technologies.[32] Lovins's ideas about energy first reached a broad audience as a result of his 1976 *Foreign Affairs* article, "Energy Strategy: The Road Not Taken?," which he expanded into a book in 1977. In the book, Lovins flipped prevailing dogmas about energy upside down: rather than assuming that greater wealth resulted from greater energy consumption, he asked if there is a point at which using more is a sign of failure.[33] He defined the soft path as conservation and the pursuit of energy from solar, wind, geothermal, biofuels, and tidal power, while the hard path used fossil fuels and nuclear power. Some advocates conceived of these paths in populist terms, the former empowering the people and the latter empowering corporations.[34] By definition, Lovins argued, the soft path technologies led to a decentralized conserving society that would empower individuals and communities. A society powered by windmills and solar panels would tend toward different values than one powered by fossil fuels. Lovins's framework helped formalize the link between the energy conservation ethic and renewable alternative energy, building on and informing mainstream discussion throughout the 1970s.[35]

The ecological paradigm on energy conservation was not merely an expert conceit. Grassroots activists joined intellectuals like Lovins and Illich in rooting their calls for energy conservation in an ecological framework of thrift. Publishing handbooks, newsletters, and magazines to spread the message and practice of living with less, scores of concerned, if comparatively anonymous, citizens worked to expose the energy waste in everyday activities in order to reshape the energy consumption of American individuals and communities.[36] The Institute for Local Self-Reliance—a community group founded in an upper-middle-class neighborhood of Washington, DC, in 1973—offers one such example. The group advocated urban agriculture, local energy production, and energy conservation and published a *Kilowatt Counter* pamphlet as "a consciousness raising tool" to empower individuals and communities to reduce their energy use.[37] Filled with statistics and an in-depth "energy awareness" questionnaire, the pamphlet took readers on an energy tour of their homes and private lives to reveal the many forms of thoughtless consumption that plagued modern life, including dishwashers, lamps, fans, air conditioners, dryers, freezers, and automobiles. It framed everyday habits in the context of population growth and resource depletion, noting the responsibility of individuals to act on their new knowledge, in part by using energy as a metric for everyday decisions about whether to drive or bike, use the dryer or hang clothes on the line. Clearly, for many in the grassroots, the burden of conservation rested on the shoulders of individuals.[38]

In addition to raising awareness, a number of research and community groups formed to experiment with alternative energy technologies and lifestyles to chart a course toward a permanent energy conservation ethic. One of the most prominent was the New Alchemy Institute on Cape Cod, Massachusetts, an alternative technology research collective established in 1969. Members of the collective worked on a small farm to develop human support systems derived from nature that used minimal amounts of renewable energy and fostered self-reliant communities.[39] Through books, volunteering, education programs, and a journal documenting their work, the group worked to spread the knowledge and practice of ecological home design, renewable energy use, and sustainable agriculture. The New Alchemists, as they were called, designed and built self-sufficient dwellings (Arks) and pioneered small-scale aquaculture, blending technological sophistication with what cultural critic Sam Binkley has described as "an anachronistic longing for the past, expressing a nostalgic, romantic appeal to a preindustrial bond with nature."[40]

The New Alchemy Institute exemplified the ways in which the ecological paradigm's vision of a post-carbon society combined the values of the energy conservation ethic with alternative energy technology. Members understood their work as part of a monumental transition from "an international, post-industrial city to a planetary, meta-industrial village," which entailed "a shift from the consumer mentality to contemplative values" and from the domination of nature to an "ecological mentality of symbiosis."[41] Cofounded by Canadian biologist John Todd, the goal of the institute was to develop technology for a decentralized, solar-powered agricultural system to preempt the predicted fuel scarcities of the 1970s.[42] Reading H. T. Odum in 1973 convinced Todd that overreliance on carbon energy explained the ongoing crises of food, population, and pollution.[43] By the early eighties, the group was claiming that renewable energy had the potential "to provide the foundation for new and more equitable societies."[44] Members believed that conservation and renewable energy would transform American society toward an ecological ethic that united sustainable thinking and doing. Although they embodied a technoscientific approach to environmental problems, the New Alchemists fell far outside the mainstream and so gained relatively little traction. Mainstream views of energy conservation imagined the energy conservation ethic much differently.

The Nationalist Paradigm for Energy Conservation

In tracing the emergence of the modern environmental movement, historians tend to stress the ways in which it diverged from earlier manifestations of environmental concern. The utilitarian conservationism of Gifford Pinchot recedes into the background after the Progressive Era in most conventional narratives, giving way to a more holistic postwar environmentalism focused on protecting "the environment," as well as the systemic, larger-scale issue of "survival" that such concern entailed, including human overpopulation, nuclear contamination, and the integrity of ecological systems.[45] Although this narrative has merit, it elides the resilience of conservationist thought—a resilience that becomes evident when considering the nationalist paradigm on energy conservation in the 1970s, which sought to manage resources in the interest of national power, independence, and efficiency.

In stark contrast to the ecological paradigm, the more mainstream nationalist paradigm framed the energy conservation ethic as a temporary strategic adaptation to support US national security and geopolitical power, as well as to stabilize the global economic system that the oil embargo of 1973 and 1974 had shaken. Petroleum had long been linked to US national security, a concern that informed, for instance, the creation of mandatory import quotas in 1959, which restricted imported oil to a certain percentage of US production to prevent dependence.[46] The 1973–74 oil embargo starkly reinforced this link, contributing to a sense of limits and decline in US power and driving energy conservation nationalists to see reduced energy use as a way for the United States to adjust to new geopolitical and energy realities. Conservation would also buy time for the nation to develop its own traditional and alternative energy resources. Those resources would be vital to support continued economic growth and geopolitical resurgence. Far from signaling a critique of petromodernity, this embrace of conservation aimed to revitalize it and the global order that it had created.[47] Perhaps unsurprisingly, this nationalist view proved predominant among federal government officials, business leaders, and media commentators.

Whereas the focus of the ecological paradigm oscillated between structural concerns and the values of individuals, the nationalist paradigm consistently laid significant blame for the nation's energy woes on its "hoggish" citizens.[48] By the mid-1970s, citizens who had been encouraged and cajoled into becoming mass consumers since the end of the Second World War, embracing an elaborate energy regime of electric lighting and appliances, automobiles, and expansive suburbs, were routinely chastised as impious squanderers of US resources. *Forbes*, for instance, welcomed oil shortages in 1973 (without a hint of irony) as a way to compel consumers to "exercise restraint in our zooming, substantially wasteful consumption of energy."[49] US oil companies spent millions of dollars scolding Americans for becoming the eager consumers that the industry had once encouraged them to be, including Atlantic Richfield, which called for a national "energy diet."[50]

This discourse of individual and collective wastefulness used religious language to frame the conservation ethic as a redemptive act to atone for the unpatriotic sin of energy gluttony. In this, it echoed popular condemnations of the 1970s as a "Me Decade" characterized by a pervasive "culture of narcissism" in which the impulses of the 1960s turned inward.[51] Members of the Ford administration thus often referred to Americans as "profligate" consumers—language that implied a moral failing but held out the possibility of national redemption through patriotic frugality to regain "full control" of America's "national destiny."[52] Jimmy Carter, who straddled the ecological and nationalist paradigms, often went further by linking the energy crisis to a deeper spiritual crisis that required the recovery of older values of sacrifice and faith in order to "seize control again of our common destiny."[53] In this configuration, energy conservation was necessary to redeem a decadent nation of narcissists whose consumption was a symptom of cultural decline. It was the recovery of a lost ethic of frugality that had once made the nation strong. At times these critiques echoed those of the environmental movement, but they were intended and applied for different purposes.

The deep "energy dependency" wrought by wasteful consumption preoccupied elite institutions and experts in the 1970s who discerned a threat to national power. The Central Intelligence Agency (CIA) put it at the top of the list of urgent geopolitical threats to American power, alongside population growth, food scarcity, terrorism, and nuclear proliferation.[54] The agency claimed in a 1977 report that energy dependence rendered the United States and its allies vulnerable to future embargoes, as well as to balance-of-payments deficits and strained security relations should Western Europe and Japan become competitors for scarce global oil supplies. Perhaps worst of all, the report warned, America's unrestrained consumption threatened to undermine its credibility as leader of the postwar economic order. If its energy consumption continued to rise, the United States could become a "scapegoat" for global oil scarcities.[55] The CIA thus advocated energy conservation as the most expedient and efficient way to reduce the need for imported oil, framing the issue as "essentially a problem of domestic political commitment and will."[56] US hegemony seemed to depend on imparting energy's national security significance to motivate action.

Strategic and geopolitical arguments for energy conservation motivated the decade's presidential administrations and key experts. As Frank Zarb, Ford's "energy czar," put it: "What is essentially at stake is the economic balance of power achieved by the Western world over the last century and a half" as well as "the restoration of American dominance in setting the goals and establishing the price of energy."[57] The Nixon administration's main response to the oil embargo hinged on the idea that energy self-sufficiency through conservation and production was essential to the preservation of America's "independent role in international affairs."[58] That the response was dubbed "Project Independence" thus proved apropos, and the nation's approaching bicentennial lent it additional cultural weight. Gerald Ford continued in the same vein, characterizing the crux of the conservation issue as the restoration of "American control of the American economy."[59] He, like Nixon, encouraged energy conservation to buy time to restore the "surplus capacity" of energy that historically had enabled the United States to shape global oil markets.[60] The Carter administration called its legislative efforts to encourage frugality and produce viable alternative energies the National Energy Security Act.[61] Environmental NGO the Worldwatch Institute likewise posited sustainability as "the key to national security" in part through the adoption of an energy conservation ethic wherein "people find energy saving in their own interest."[62] It seemed clear to these and other policy leaders that high energy consumption threatened US security and that conservation would be vital to dampening potentially divisive resource competition.

The strategic value of energy conservation explains why officials like Secretary of State Henry Kissinger could opine about conservation while placing hope in the potential of thermonuclear fusion to "make available abundant energy from virtually inexhaustible resources."[63] For a realist like Kissinger, global oil had become a liability to US power, necessitating conservation until alternative energies or American oil could be brought online. His view carried over into the Carter administration. The stated objective of Carter's 1977 National Energy Act, the brainchild of former CIA director and secretary of defense James Schlesinger, was to "reduce dependence on foreign oil and vulnerability to supply interruptions" and to develop "renewable and essentially inexhaustible sources of energy for sustained economic growth."[64] Until such time, conservation would have to mitigate America's "excessive" and "intolerable" dependence on OPEC.[65]

The dominant, nationalist discourse of energy conservation, then, revolved around the restoration of US geopolitical power and abundant energy consumption, rather than the goal of permanently reduced energy consumption that animated the ecological paradigm. The capacities of petromodernity, as well as its sociopolitical commitments and ecological consequences, remained essentially unchallenged. Individual citizens now bore the brunt of the blame and were often cajoled into becoming, if only temporarily, conserving citizens willing to forgo many of the conveniences of carbon capitalism by internalizing an ethic of energy conservation.

"Don't Be Fuelish": Chastising US Energy Consumers

Although federal efforts to manage US energy supplies in the 1970s were varied and fraught with disagreement, interest in energy translated into policy. Before the oil embargo, Nixon adjusted existing oil price controls and introduced allocation measures to distribute access to oil; afterward, his administration proposed a variety of initiatives, including approval for the controversial Trans-Alaska Oil Pipeline, yearlong daylight saving time, a 55 mph speed limit on highways, new petroleum allocation powers for the president, and requests for voluntary assistance from the public.[66] Through Project Independence, Nixon endorsed national conservation measures and "alternative" energy production, though the

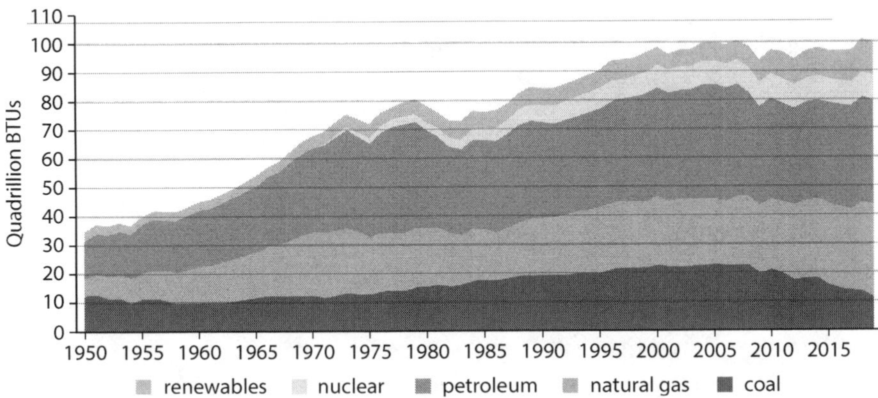

Figure 6. US primary energy consumption by major sources, 1950–2019. Source: Data from US Energy Information Administration.

latter mostly meant building hundreds of nuclear power plants by the year 2000.[67] Ford also pursued production, advocating nuclear power plants and opening up more land and water to oil development, especially on the Outer Continental Shelf. Carter's first energy plan, drawn up by James Schlesinger, stressed energy efficiency and conservation through taxes and higher oil prices with a promise to return "windfall" profits to consumers.[68] After the House and the Senate diluted the plan, Carter shifted to production via gradual oil price decontrol and federal investment in alternative energies, including oil shale, geothermal, and solar, through an "Energy Security Corporation," a slight departure from the emphasis of previous administrations that drew the ire of economic conservatives.[69]

In the midst of this focus on energy production and deeply partisan energy policy debates, Congress managed to pass important energy conservation measures. The 1975 Energy Policy and Conservation Act introduced fuel efficiency standards and labeling for automobiles and "energy efficiency ratings" for household appliances. The 1977 National Energy Act built on this success by developing minimum efficiency standards for appliances and tax credits to encourage homeowners to renovate their homes to be more energy efficient.[70] These achievements combined with the nation's economic struggles to slow the rate of energy consumption growth for a few years, but they were ultimately minor victories in the effort to use less energy. Continued reliance on automobiles and other infrastructures of petroculture meant that growing population and economy translated into growing total energy use (fig. 6).

Most legislative efforts to foster energy efficiency relied on the assumption that meaningful reductions in energy use boiled down to "countless decisions by individual citizens."[71] Regulating the efficiency of engines and appliances was important, but ultimately the federal administrations of the 1970s shied away from dictating energy use. They would instead have to rely on voluntary changes in behavior. Worried that too many people failed to grasp the urgency of the energy crisis, these administrations used public education campaigns to inculcate an energy conservation ethic in US citizens as a key way to achieve energy independence. Although their rhetoric and aims sometimes overlapped with that of the ecological paradigm, government-sponsored education campaigns aligned with the nationalist paradigm, framing the energy crisis as a problem of individual habits and so sought to mobilize individual citizens to discipline their everyday energy use in order to regain national vitality.

Energy-conservation campaigns tried to encourage voluntary participation by convincing *all* Americans that their individual actions mattered. As the Federal Energy Office argued in March 1974, "Effective energy conservation depends upon millions of wise individual decisions about lifestyle and buying habits. Individuals must come to understand that they are a part of the problem and that their actions are a part of the solution. In short, we must develop an 'energy ethic.'"[72] Zarb's "Energy Sense" column directly addressed consumers along these lines, advising that energy could be saved by small adjustments: avoiding rush hour traffic, using minimal water for cooking, turning off unused home lighting, and moving furniture away from heating vents. The inclusion of energy news and statistics alongside his advice embedded these mundane tasks in a larger context that imbued them with geopolitical significance.[73]

One of the most pervasive federal campaigns for energy conservation was the Ford administration's "Don't Be Fuelish" television and radio PSA campaign, which framed energy conservation as a moral choice between frivolous private wants and vital public needs. The slogan's wordplay implicitly linked energy scarcity to the actions of individuals who consumed without care or foresight. Created by the venerable advertising firm Cunningham and Walsh, the ads premiered during the Super Bowl and appeared on radio and television throughout 1974 and 1975.[74] They used authoritative entertainment and sports celebrities, including actor George C. Scott and Miami Dolphins head coach Don Shula, to warn that failure to conserve could lead to social disintegration. In the Super Bowl spot, Shula warned that fighting the energy crisis was "no game" and concluded with an appeal to self-interest: "If we all help, we'll really be helping ourselves." Scott implored viewers from an empty classroom that they would have to "give up something": either their "lavish" energy consumption or public amenities like schools.[75] In blaming individual consumers, the ads ignored the history of state and corporate efforts to integrate citizens into an energy-intensive modernity by encouraging the purchase of suburban homes, automobiles, and appliances.[76] For the first time since the Second World War, American citizens and businesses were told to sacrifice the possessions and activities that were by now woven into the fabric of everyday life.

The Carter administration went deeper, publicly framing energy conservation as a project of social and spiritual renewal that required individual sacrifice. Carter told a Kansas City audience in 1979, "Just as the energy shortage has forced us to face our deepest fears and divisions, so our goal of an energy secure America will help us to rebuild our strength and our confidence as a people." He famously compared the energy challenge to war and worried aloud about the nation's ability to unite in the face of it. In his view, dependence on oil imports signaled a weakened collective will, which stood in marked contrast to the past when the United States "sent more oil out of the country than we brought in." In order to reclaim this lost vitality, Carter called on "every single American [to] stop wasting energy." Conservation was, he said, "a matter of patriotism and of commitment."[77]

Carter's administration designed several education initiatives to encourage Americans to sacrifice. Its Council on Energy Efficiency enlisted prominent citizens to promote conservation in their public appearances and publicly recognized individuals and organizations making outstanding contributions to fuel conservation.[78] The Advertising Council developed an optimistic media campaign titled "Keep It Up America" to congratulate citizens on their conservation efforts. The campaign saturated the airwaves, reaching an estimated 120 million Americans over the summer of 1980.[79] The administration's other initiatives included carpool education, PSAs, and demonstrations, as well as energy conservation brochures and films for homes, farms, and motels. These materials all sought to "convince every single American that he or she, as just one individual, can take action that will truly be significant."[80] Indeed, everyone had a moral responsibility to do so.

In an effort to shape the nation's future leaders (who might also convert their skeptical parents), the federal government tailored its conservation message for children and teens.[81] The National Science Teachers Association thus developed energy-conservation school curriculum for the Energy Research and Development Administration in 1976–77 intended to make conservation "meaningful" to young people.[82] To that end, new first-grade curriculum introduced the concept

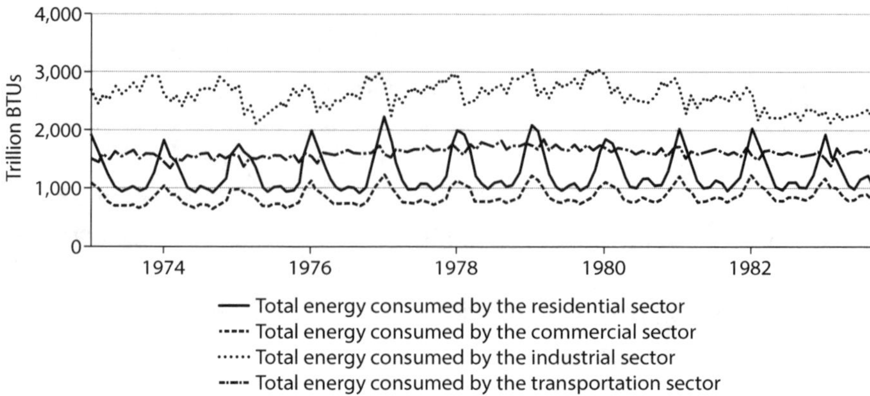

Figure 7. US energy consumption by sector. Source: Data from US Energy Information Administration.

of energy through heat and motion, stressing how it linked each child to the wider world. One lesson asked children to stretch and move to produce heat energy and then asked each child to close his or her eyes and "imagine what energy feels like."[83] After instilling an embodied sense of what energy is and how it connected children to nature, later curriculum taught energy-conservation practices and the history of human energy use. The Young Energy Savers (YES) curriculum suggested activities and games to instill energy thrift, including making a shopping list to save gas by shopping more efficiently, using buckets instead of running water to wash up, and doing community service to eliminate energy waste.[84]

These and other education efforts, which included oil company campaigns against energy waste, achieved limited results.[85] By the late 1970s, polls suggested that a majority of Americans believed that their personal conservation efforts could help to redress energy shortages, and a majority remained concerned about energy conservation at the household level in the early 1980s.[86] But many also continued to favor the development of "hard path" energies such as coal, oil, natural gas, and nuclear power while, perhaps more tellingly, total US energy consumption continued to rise.[87]

Many of the figures in the ecological paradigm who advocated a renewed social order rooted in alternative energy technologies lacked the institutional power of the nationalist paradigm and its emphasis on linking the everyday choices of individual citizens, including children, who might leave room lights on, to US oil dependence. As a result, residential consumption (fig. 7) occupied a disproportionate amount of public discourse, eventually generating a backlash; by the time

Ronald Reagan rode into office, public discourse had turned to frustration with what the *New York Times* called "fiddling with the thermostat."[88] Nevertheless, in oil price decontrol, the Ford and Carter administrations, pro-market experts, and oil companies believed that they had a way to ensure an energy conservation ethic that did not rely on voluntary behavior. They saw the free market as an effective tool to discipline energy consumption. Crucially, important environmentalists agreed.

"Market Forces": Disciplining US Energy Consumers

By the mid-1970s, policy makers and experts heralded oil price decontrol as a deregulatory solution to the energy crisis that would curb energy consumption and stimulate production at the same time. The federal government had a long history of intervening in the oil market, having, among other things, imposed oil import quotas under Eisenhower and oil price controls under Nixon. In the wake of the oil embargo, experts complained that these quota and pricing regulations had distorted the oil market by keeping US prices artificially low, and in so doing had removed incentives for domestic oil production and energy conservation.[89] Decontrolled prices, the argument went, would produce a "conservation effect" by raising rates. Higher prices would compel users to manage their consumption in the short term and generate profits to fund US energy resource development over the long term.[90] Economists and key members of the Ford administration thus sensed in the energy crisis an opportunity to tie the conservation ethic they sought to inculcate market forces, essentially turning widespread agreement about the need for energy conservation to an endorsement of a free market economy.[91]

Decontrol formed the core of energy policy in the Ford administration. The Federal Energy Administration reasoned that higher oil prices would force individuals to use energy more sparingly and boost domestic oil supplies. Higher production would enable the United States to be effectively energy independent by the 1980s, thereby reducing vulnerability to future embargoes and preserving America's geopolitical leadership.[92] To cushion the impact of inevitable price spikes, the administration considered removing fees on imported oil, but it warned repeatedly that refusing to let oil prices rise would only worsen the nation's vulnerability and decrease its capacity to produce and profit from its own undervalued oil resources. Steered by pro-market advisers William Simon and Alan Greenspan, the administration decided that Project Independence should rely on the market as much as possible, including the "total decontrol as soon as possible" of all energy prices.[93] Indeed, most influential voices within the administration believed that "market forces" would conserve substantial amounts of energy in the short term, obviating the need to rely on appeals to voluntarism, patriotism, and sacrifice.[94] The free market would produce any desired behavioral changes by disciplining consumers. Although Ford had to compromise with a resistant Democratic Congress, phasing out controls much more slowly than he wanted, support for decontrol grew as the decade wore on.[95]

One of the animating ideas in decontrol, that Americans should be paying more for oil (as well as other forms of energy) to encourage conservation, resonated with mainstream, reformist environmental groups, who also began to support decontrol as an expedient way to achieve their goals.[96] Although Meg Jacobs has demonstrated the fiscally conservative impetus behind decontrol, it is important to note environmentalists' support for it insofar as they, and much of the ecological paradigm, took higher energy prices to be essential to a conserving society.[97] Thus, on March 24, 1979, the Conservation Foundation, the Sierra Club, the League of Women Voters, the Natural Resources Defense Council, and the Worldwatch Institute endorsed oil price decontrol in the *Washington Post*, noting that higher prices would dispel the idea that "profligate consumption is cheap."[98] Denis Hayes, one of the key organizers of the first Earth Day, argued for the decontrol of *all* energy prices (natural gas, oil, and electricity), noting that paying their true cost would reduce consumption and render alternative energy technologies economically more attractive. These arguments echoed the claims of many proponents of the ecological paradigm on energy conservation, who had been arguing since the early 1970s that energy resources had to be valued more highly to account for the ecological costs of their production and consumption, and to generate a sustainable society.[99] But whereas prominent thinkers in the ecological paradigm advocated taxes, quotas, alternatives to money, and rationing to value energy resources differently, mainstream environmental groups made peace with the market to discipline consumption, as long as it included some sort of windfall profits tax.[100] As ecologist Kenneth Watts testified to Congress in 1976: "A rift has developed between the environment movement . . . and the consumer movement."[101]

In so doing, these groups agreed with the Ford Foundation, which had just abandoned the zero growth conclusions of its influential report *A Time to Choose* (1974). In its 1979 report, the foundation advocated market-based price incentives to encourage energy conservation as well as the decontrol of all oil and gas prices. It claimed that "total decontrol of prices paid to domestic producers . . . would have constructive effects on patterns of energy consumption, on choices between imported oil and domestic energy sources, and on long-run investment choices in both energy production and consumption."[102] Although many Americans resisted the idea of higher prices in the era of stagflation, the public also seemed to favor decontrol by mid-decade, though their support tended to emphasize the promise of stimulating domestic energy production rather than the prospect of demand-dampening prices.[103]

Although Jimmy Carter dismissed "immediate and total decontrol [as] disastrous for our economy and also for working American families," he eventually mandated gradual oil price decontrol.[104] On April 5, 1979, he ordered the cessation of domestic crude oil price and allocation controls by September 30, 1981.[105] The move blended Carter's fiscal conservatism and his environmentalism. Carter's rhetoric matched environmental groups, ecologists, and others who saw decontrol as a step toward valuing energy at its true "replacement value" to foster more conscientious consumption.[106] But it also echoed free market economists and oil companies by framing the decision to enact phased-in decontrol as the ending of "excessive federal government controls."[107] Upon signing phased decontrol into law on June 1, 1979, Carter praised the policy as a panacea for the interlocking energy problems that had plagued the decade: increasing domestic oil production, cutting consumption, reducing vulnerability, and improving balance of payments.[108] When Reagan ended oil price controls during his first week in office in 1981, he nodded to "prudent conservation" but showed more concerned to stimulate "vigorous domestic production."[109] His administration claimed that the American people, participating in a free market for oil, could now choose for themselves how much energy was appropriate to consume.[110] As oil prices fell thereafter, *Harper's* declared the end of the energy crisis.[111]

Conclusion

The energy conservation ethic of the 1970s aimed to reduce US energy consumption by demonstrating the depth of national dependence on fossil fuels and encouraging consumers to internalize frugality in their everyday uses of energy. Its popularity, which demonstrates the persistence of conservationism, grew out of a loose coalition of environmental advocates working to create a sustainable society and of more mainstream conservation nationalists looking to restore US geopolitical power by using fewer energy resources from foreign powers and boosting domestic energy production. These ecological and nationalist paradigms both advocated valuing energy more highly, producing discourses that legitimized austerity and free market oil pricing as a way to discipline energy consumption and spur domestic energy resource development. Crucially, however, the mainstream views of this solution to the problem of energy did not entail imagining an alternative future but instead sought to use the market to discipline citizens temporarily while promising abundant energy in the future, largely through the intensified exploitation of American lands and seas. The denouement of America's vibrant national preoccupation with energy conservation thus marked a triumph for the nationalist paradigm, which tracked with the broader rightward turn of US political culture and re-entrenched the role of oil in US social and economic life at the end of a decade in which another future seemed possible. It set the stage for the oil wars of the 1990s and 2000s and the ongoing power of market-based approaches to our most pressing environmental issues: climate change and energy transition.

The turn toward market pricing killed the conservation ethic that environmentalists had hoped to inculcate. Recall that for Aldo Leopold a land ethic entails "an internal change in our intellectual emphasis, loyalties, affections, and convictions."[112] By arguing for market prices, mainstream environmentalists decided to rely on self-interest to achieve their conservationist goals rather than the inculcation of new values, attitudes, and loves that might bring forth new relations between humans and the earth and between communities. Emissions regulations, energy switching, and market pricing may have achieved modest deductions in per capita energy use, but they did so *within* the technocratic order of centralized power and energy-intensive machinery that thinkers like Jacques Ellul, Ivan Illich, and Wendell Berry had been warning about—an order dominated by the logic of machines to "get bigger and more elaborate."[113] The broader environmental movement may have changed some attitudes about pollution, recycling, and "protecting" the earth, but when it came to energy, petrocultural values survived the energy crisis era relatively unscathed.

"Put Your Foot on the Pedal"

Contesting Conservation in Seventies Car Cinema

> Why all this speed and trouble and risk?
> —BEAUTY (FARRAH FAWCETT), *THE CANNONBALL RUN*

Not everyone embraced the energy conservation ethic. Despite expert enthusiasm for conservation, many Americans continued to doubt the necessity of adjusting their consumption habits. High oil prices and energy conservation measures seemed unnecessary inconveniences at best or abuses of power at worst. When in 1977 Jimmy Carter proposed that Americans treat the energy crisis as the "moral equivalent of war" to inspire a national conservation effort, critics discerned a threat to the postwar ideal of unlimited automobility rooted in cheap and easy oil consumption.[1] As noted by one commentator who mockingly dubbed Carter's proposal "MEOW," many Americans stood at the "ready to fight anybody who tries to cut down on their motoring pleasure, energy crisis or not."[2] Given the danger that the energy crisis and conservation measures posed to what Matthew T. Huber has called the "deeply felt visions of freedom and individualism" at the heart of US petroculture, acts of resistance cropped up across the country.[3]

Resistance took many forms. In December 1973 and February 1974, thousands of mostly nonunionized truckers ("owner-operators") organized slowdowns and roadblocks primarily in the northeastern states to protest rising fuel prices and the 55 mph speed limit intended to conserve fuel. The truckers claimed that their vehicles operated more efficiently at higher speeds, so having to drive 55 mph burned through more fuel and prevented them from working at a profitable pace. To broadcast their concerns and compel the government to act, truckers organized to slow down traffic, refused to deliver loads, or stayed home. Similar protests flared up sporadically across the country for the duration of the energy crisis, sometimes involving acts of violence and intimidation.[4] Most infamously, frustration boiled over into violence in June 1979 when a protest staged by truckers and residents of Levittown, Pennsylvania, descended into nearly two days of rioting.[5] Needing to drive fast and far to survive, independent truckers could not abide higher prices and tighter regulations, nor could the emerging postindustrial service economy that they served.

If trucker protests decried the economic burden of lower speeds and higher prices, other developments expressed contempt for government efforts to regulate American motoring. Journalist Brock Yates created the Cannonball Baker Sea-to-Shining-Sea Memorial Trophy Dash in 1971, an illegal cross-country road race from New York to Los Angeles to celebrate America's Interstate Highway System and to protest the stricter traffic laws then coming into effect. He resented the EPA's new emissions standards and the threat that "legions of nanny state bureaucrats" posed to the freedom of the road.[6] When the Nixon administration introduced the 55 mph speed limit in 1973, the Cannonball race took on even greater significance as a rebellion against government overreach.[7] On its August 1975 cover featuring Cannonball participants, *Car and Driver* called the 55 mph law "the dumbest law since prohibition."[8] For the Cannonball drivers and their fans, energy conservation threatened to suffocate American motoring and the ethos of freedom associated with it. As an act of resistance to state authority over automobility amid the energy crisis, the Cannonball road races suggested that US petroculture would not go quietly into a frugal, alternative energy future. And like the trucker protests, the road race inspired popular film cycles that voiced the energy conservation resentment of millions of Americans.

This chapter pivots from energy conservation efforts to populist backlash against energy conservation through an analysis of popular culture. Popular films engaged with energy matters in the 1970s, often using cars and carbon-powered mobility to explore freedom, state power, apocalyptic anxieties, and escapist fantasies.[9] I argue that the narrative and aesthetic content of these car films, as well as their popularity, demonstrate a growing rejection of the conservation ethic by the late 1970s. Generally adopting a "petro-populist" position on energy conservation, many car films posited uninhibited automobility as the sacrosanct right of the American people while celebrating fossil-fueled mobility as a form of resistance to state control over energy consumption. Featuring working- and middle-class men who rebel through fast and often reckless driving, these films mocked expert eulogies for carbon civilization and valorized cars and driving as a source of identity, independence, and power. Yet in their oscillation between celebrating the symbols of US petroculture (i.e., the open road) and expressing anxiety about its viability in images of violence and decay, car films suggest cultural ambivalence about the energy crisis, conservation, and the American future. Thus, popular cinema reveals a postwar energy imaginary undergoing a crisis and a reformulation in which freedom of the road stands in for economic freedoms writ large threatened by state regulation.

There are four parts to this chapter. The first covers the portion of popular seventies cinema that signaled a crisis in the energy imaginary of the postwar United States by critiquing oil. Echoing certain environmentalist and counterculture views on energy, films such as *Five Easy Pieces*, *Three Days of the Condor*, and *King Kong* imagined oil as a source of alienation, corruption, and ecological destruction. These are the kinds of cultural imaginaries that undergirded critiques of US energy consumption and called for change. The chapter's remaining sections explore emerging cultural resistance to a conservation ethic in cinematic representations of automobility and its associations with core cultural values like freedom and individualism in "car films."[10] The second part of the chapter examines how nostalgia for US car culture in *American Graffiti* and *Damnation Alley* celebrated motoring and tied it to supposed cultural innocence. The third part explores the petro-populist side of car films, which linked automobility to personal and political freedom that must be protected from state regulation in *Convoy* and *White Lightning*, representatives of the "trucker" and "southern" car films. The final section explores defiance and ambivalence in car race and chase films like *Dirty Mary Crazy Larry*, *Death Race 2000*, and *Cannonball Run*. As bemused critics reported, audiences took great pleasure in watching cars get dented, dismantled, crashed, and spectacularly blown up. Even as films about cars, driving, truckers, and the open road articulated popular resentment of the conservation ethic, the nihilistic and wanton destruction of cars on the big screen belied an undercurrent of anxiety about the future of US automobility. This chapter's analysis of popular US cinema thus reveals links in cultural discourse between the energy crisis and popular suspicions of state power over markets and individuals that is at the core of neoliberalism. It reveals the retrenchment and neoliberal reformulation of the cultural imaginary of oil in response to the energy conservation ethic in the 1970s.

The Threat of Oil

In the late 1960s, counterculture and environmentalist critiques of America's postwar energy imaginary, such as Edward Abbey's *Desert Solitaire*, began to seep into popular culture. This development is evident in the spate of films critical of oil's social and political effects, which implicated oil culture in generational alienation, political corruption, and ecological destruction. American road movies, for instance, shifted from equating automobility with countercultural subversion of a conformist society to a more nihilistic stance on the alienation inherent in US petroculture.[11] For instance, the counterculture classic *Easy Rider* (1969) posited freedom from an oppressive society *through* automobility as it followed two friends on an aimless ride across the American Southwest.[12] But other films, like *Two-Lane Blacktop* and *Vanishing Point* (1971), used the open road to explore how empty and meaningless American life had become. As critic Greg Ford wrote of *Two-Lane Blacktop*, the film's pointless race from Los Angeles to Washington, DC, was "a kind of aimless, listless, rootless game that nobody wins, played out by people who are sadly convinced that there's nothing better to do."[13] Similarly, Terrence Malick's "gasoline opera" *Badlands* followed young lovers (Sissy Spacek and Martin Sheen) on a doomed, nihilistic joyride across the plains of the Midwest after committing patricide.[14] Often emphasizing the rootlessness of its malicious protagonists, the film warns about the consequences of an alienated generation of young Americans.

The most iconic film to use mobility and oil imagery to critique America was Bob Rafelson's 1970 hit *Five Easy Pieces*, which the Library of Congress selected in 2000 for preservation in the National Film Registry. Protagonist Bobby Dupea (Jack Nicholson) was the quintessential "unmotivated hero" of a cycle of early seventies antiheroes, and audiences enthusiastically embraced him as an expression of their discontent.[15] The film introduces Bobby, the estranged son of a wealthy family from the Pacific Northwest, trying to live an authentic life away from his privileged and stultified upbringing. Bobby works in the oil fields of Southern California, driven by what one critic called "the romantic and condescending notion that the working classes are the sole possessors of a mysterious vitality denied the uptight middle classes."[16] He dates a simpleminded waitress, Rayette (Karen Black), and befriends working class people over whom he feels superior. Dissatisfied with his attempt at working-class authenticity, Bobby visits his sister in Los Angeles and learns that his estranged father has suffered a stroke. The rest of the film follows Bobby's trip up the coast to his family's island estate outside of Seattle, where he tries to reconcile with his father but fails to overcome his contempt for his family's stilted life. In the end, Bobby abandons Rayette at a gas station and hitchhikes on a logging truck headed to Alaska.

Five Easy Pieces uses oil infrastructure and mobility imagery to express Bobby's alienation and dissatisfaction. Unaligned with any cause, idea, or person outside of himself, Bobby's estrangement is implied visually in the deserted, uninhabited landscapes of oil extraction. The oil derricks are not a source of community or class consciousness, but rather of difficult and lonely labor. If this is working-class life, its daily grind is unappealing and uninspiring, much like the fabricated petroculture that exploits it. Moreover, mobility and immobility imagery frame Bobby's estrangement from family and society. His family is marked by immobility: the family estate is cut off from the world on an island that is only accessible by ferry, Bobby's brother wears a neck brace, and his father has suffered a severe stroke. The estate bears little evidence of twentieth-century car culture, more closely resembling aristocratic grounds before the advent of the automobile. Bobby, in contrast, is healthy and mobile and cannot bear to be still. In an early scene eerily foreshadowing the frustrations of gas lines in late 1973, Bobby and his coworker get stuck in a traffic jam on their way home from work. Bobby hops out of his car and lashes out at the other cars blocking the road before jumping on the back of a truck to play a piano to the hive-minded drivers confined to their cars. Later in the film, Bobby confesses to his father: "I mean, I move around a lot because things tend to get bad when I stay. And I'm looking . . . for auspicious beginnings, I guess." Constantly on the move, Bobby views his family's life as stolid and inauthentic.

Five Easy Pieces is ultimately ambivalent about Bobby's constant movement. Promotional material drew heavily on the symbols of US petroculture, picturing Bobby staring dully into the distance framed by oil derricks and the tagline, "He rode the fast lane on the road to nowhere," suggesting that his restless motion is a source and symptom of his alienation. The film's lingering final shot of a lonely service station reinforces this reading. In the final scene, Bobby hitchhikes on a logging truck headed north, abandoning Rayette while she gets coffee inside the service station. Bobby's choice for more aimless wandering on the road raises his disaffection to the status of moral irresponsibility, rather than countercultural rebellion. Furthermore, this service station setting is a far cry from the postwar modernist imaginary of progress and freedom. It is, instead, lonely and isolated, an image of social disconnection and the disintegration of community, where Bobby commits the film's final antisocial act. With a Gulf Oil sign hanging prominently in the middle of the frame (fig. 8), the service station symbolizes the film's

Figure 8. The final shot of *Five Easy Pieces*, after Bobby leaves his girlfriend, Rayette, for the open road to Alaska. This roadside stop is one of many alienating artifacts of US petroculture in the film. Source: *Five Easy Pieces*, BBS Productions, 1970, Criterion Blu-ray, 2015.

themes of alienation and anonymity; it is a "no place," created by the constant movement of culture that creates spaces of separation and meaninglessness.

In the 1970s, popular culture also articulated growing suspicion about the corrupting influence of oil that informed populist accusations of collusion between "Big Oil" and "Big Government." Sydney Pollack's box office hit *Three Days of the Condor* (1975), for instance, decried the oil-driven duplicity of the CIA as a threat to democracy.[17] The film follows Joe Turner (Robert Redford), a CIA researcher on the run after an assassin murders everyone in Turner's office to stop him from digging deeper into the machinations of the agency. In a mise-en-scène dripping with paranoia and distrust, Turner uncovers a rogue group within the CIA intent on invading the Middle East to secure oil for the United States. Although Pollack preferred to see the film as a spy thriller rather than a political statement, its "cool, amoral style of conspiracy" resonated with the ongoing revelations surrounding Watergate and the CIA at the time of its release, as well as popular fears of oil shortage and national dependency.[18] When Turner asks Deputy Chief Higgins if the CIA plans to invade the Middle East, Higgins retorts that the American people "just want us to get [oil] for them." The political power afforded to the state by its ability to supply Americans with abundant oil is too tempting, the film says, providing incentives for corruption and war.[19]

An explicitly environmentalist perspective on the threat of ecological destruction inherent in oil drove the plot of Paramount's remake of *King Kong* (1976), which succeeded at the box office despite a generally frosty reception from critics.[20] In an effort to update the story, director John Guillermin and writer Lorenzo Semple Jr. used oil exploration, rather than filmmaking, as the motivation for the journey to King Kong's mysterious island. The film begins with a Petrox Petroleum crew preparing for an expedition to the island, where illegally obtained NASA satellite information says there is untapped oil. But, as critic Richard Eder noted, *King Kong* was "unreservedly pro ape."[21] The greedy oil company, as personified by Charles Grodin's sniveling expedition leader, is the clear villain. The audience is encouraged to cheer for Princeton zoologist Jack Prescott (Jeff Bridges) and Kong. Grodin's irrepressible desire for money drives the plot and tramples all other interests. He heedlessly pursues the oil pooling on the island's surface, and when he discovers that it is worthless, he tries to turn a profit by bringing Kong to New York City as a Petrox Petroleum mascot and the star of a live "beauty and the beast show." Kong escapes his cage in a rampage to save Dwan—the blonde woman that he desires—and eventually heads to the World Trade Center towers that resemble the mountain peaks of his island, where he is killed. In the tragedy of King Kong, the film criticized feckless, oil-profiteering businessmen and their abuse of centralized power.[22]

Besides linking the pursuit of oil and corrosive greed, *King Kong* critiqued the rapacity of US petroculture, echoing larger concerns about oil's environmental consequences. The trip to Kong's island is driven by the corporate exploitation of oil resources that knows no limit. Early in the film, Jack Prescott—the film's environmental conscience—foreshadows King Kong's fate by accusing Petrox of "environmental rape." Not only is the company unconcerned about the ecological consequences of extracting oil from Kong's island, it shows no compunction about exploiting Kong himself, whose expressive face, emotional attachment to Dwan, and persecution render him as a noble savage.[23] Indeed, Kong symbolizes the pure power of nature that the oil company wants. As critic Ernest Larsen noted, Kong has always been a symbol of power, but his 1970s incarnation is "much more clearly a symbol of defenceless power, the raw energy of an undeveloped resource."[24] The film repeatedly equates Kong with raw power in its action sequences, in which he effortlessly destroys human attempts to contain him. In the end, Kong literally stands in for oil: as the crowd awaits its first sight of the ape, an enormous gas pump rolls into the stadium as Grodin intones, "Ah, the power of it. The superpower. Hail to the power!" Kong is hidden under the gas pump's facade, the new mascot for Petrox Petroleum and a literal symbol of the exploitation of nature's power for human ends. Ultimately, Kong is killed atop the World Trade Center towers—potent symbols of global capitalism—and becomes a spectacle for the thronging masses that crowd around his corpse. In Kong's lifeless body we glimpse spent nature, wholly consumed—a warning about America's relation to oil and to the nature from which it comes.[25]

In its expression of ecological anxiety and opportunistic critique of capitalist greed, *King Kong* added concern about US energy resources to the decade's cycle of ecologically themed films, which tended to focus on human population and pollution. These included *Silent Running* (1972) and *Soylent Green* (1973). Most cinematic engagements with energy, however, took automobility as their core theme and articulated a very different perspective on the meaning of the energy crisis and energy conservation: as despotic threats to the individual freedoms facilitated by petroculture.

Free to Choose the Freedom of the Road

Postwar growth in private car ownership remade the landscape and economy of the United States. Roadways multiplied while the design, manufacture, maintenance, marketing, and driving of automobiles generated millions of jobs and changed how Americans lived and moved. This car culture also changed how they dreamed. As James Todd Uhlman and John Heitmann have argued, mass automobility spawned an imaginary that linked "the unrestrained capacity to move . . . with personal reinvention and self-determination."[26] Fossil-fueled mobility became a synonym for liberty, freedom, and agency, rooted in what Huber calls the "privatized command over space" that cars fostered.[27] This imaginary was a cornerstone of neoliberal political culture, making it easier to see one's life as separate from broader social forces. The energy crisis and the environmental movement helped to call this postwar energy imaginary into question, as films like *King Kong* and *Three Days of the Condor* suggest. But the proliferation of popular car films in the 1970s bespeaks the presence of competing notions about the implications of the energy crisis and conservation for American citizens. They attest to an emerging petro-populist rejection of energy conservation that resisted state control over automobility and valorized a neoliberal vision of the self as one who doggedly pursues individual liberty and entrepreneurial risk-taking.

Energy conservation discourse often hinged on questions of selfhood and individual responsibility. As we have already seen, the conservation ethic dictated that reducing energy use required the integration of laws and practices with new ways of being and feeling. Using energy was so habitual that its control required the formation of disciplined selves with new habits. By 1979, conservation nationalists and environmentalists agreed to a neoliberal détente that called for "market forces" in the oil sector to compel individual selves to curtail their energy consumption.[28] But as popular film makes clear, neoliberal notions of selfhood also emerged in populist disdain for energy conservation regulations, which were decried as the illegitimate intrusion of a nanny state on the freedom to choose how much energy one consumed and for which purposes. Populist discourse and practice, from trucker protests and illegal road races to the films that celebrated them, joined an emerging collective of anti-limits economists, pro-market political leaders, and intellectuals who celebrated such freedom of the road as the inalienable right of the American people to be protected from meddling bureaucrats. This was ironically a petro-populism that prioritized the right of individuals to choose their mode of mobility, free from the needs or constraints of the larger society. Many of the protagonists in these films were entrepreneurs who relied on their own human capital to make a living.[29] They often embodied Sam Binkley's notion of the neoliberal self as a "loose," flexible, and chosen self that emerged to cope with the dynamic instability of consumer capitalism in the 1970s, replacing the relatively fixed ways of being formed by disappearing traditions and communities.[30] Self-made *men*, car film heroes thrived on risky behavior, regarded citizenship in individualistic terms, and ignored the rule of law.[31] Such men populated cinematic struggles to maintain freedom of the road—to choose one's mode of movement—against threat of the energy crisis and the conservation ethic.

The dozens of car films that raced through American cinemas in the 1970s marked a decisive shift in popular representations of motoring in response to the energy crisis and injunctions to conserve.[32] Many 1960s films celebrated the joys and freedom of driving, but the car films of the 1970s were different.[33] They defended the freedom of the road through nostalgic celebration that gave way to defiance against state efforts to conserve energy. George Lucas's *American Graffiti* (1973), whose aesthetic dripped with nostalgia for the more "innocent" car culture of the past, and Twentieth Century–Fox's *Damnation Alley* (1977) argued for the inherent goodness of American petroculture. A second set of films defied the conservation ethic by celebrating uninhibited automobility as integral to individual freedom and agency, particularly for the alienated men of late sixties cinema who were resurrected as risk-taking (anti-) heroes in seventies car films. These films included what Derek Nystrom and Warren French call "the southern" as well as the "trucksploitation" film cycles.[34] In both, working-class male protagonists locate their identity and livelihood in their automobiles, which they use to resist and outwit incompetent bureaucrats. Then there were the car crash films that, in their repetitive and wanton destruction of cars, belied an undercurrent of working-class anxiety about the status of the automobile in the twentieth-century United States.

Petroculture Nostalgia

American Graffiti arrived at a propitious time. Released in August 1973, the wildly popular celebration of car culture played in cinemas for the duration of the 1973–74 oil embargo. As millions of Americans waited in line to buy gasoline at inflated prices and fretted about the future of the oil flows that pumped the pistons of middle-class American life, George Lucas's film offered a comforting remembrance of the innocent joys of motoring in a purportedly more stable and prosperous time. Conceived and widely recognized as a nostalgic rendering of middle-class American life before the 1960s tore the nation's social fabric, the film follows four high school friends on a late summer night in 1962 as they cruise their small California town. It is their last night together before moving on with their lives full of fitful attempts to hold on to youth and cope with the looming choices that will shape their adult lives.[35] Featuring an expensive 1950s rock 'n' roll soundtrack, authentic cars, and location shooting, *American Graffiti* labored to render an innocent depiction of Eisenhower's America on the eve of its maturation. It brought audiences into the concerns about love, career, identity, reputation, and friendship of a group of likeable teenagers on the cusp of lost innocence.

Lucas's nostalgic vision of his generation's youth revolved around automobility and the unique cultural practices that accompanied it, including its "mating rituals," cruising, and drag racing.[36] *American Graffiti* essentially romanticized and memorialized the apex of American petroculture before environmentalist and energic discourses threw its decency and viability into question. The film identifies its main characters with their cars, and the characters themselves embrace their cars as status symbols and spaces of possibility. The only exception is Curt (Richard Dreyfuss), who aspires to be a writer and cares less for cars than his peers do. For most of these kids, cruising is a cherished way of life and source of social identity. As critic Vincent Canby noted, most of the film "takes place inside, on top, underneath, or within spitting distance of cars—Fords, VWs, Chevies, Citroens, Pontiacs" and features a generation of youth defined by "their obsession with movement."[37] The garish neon lighting and the sense of constant speed captured by cinematographer Haskell Wexler offered a romantic and wistful sense of a past car culture whose future seemed uncertain, but only in retrospect. Indeed, Fredric Jameson has dubbed *American Graffiti* the first in a new genre, the "nostalgia film," for its pastiche construction addressing a certain generation with a romantic vision of its past rooted in cultural stereotypes rather than historical realism.[38]

Although some critics insist that Lucas's oeuvre should be read as a subtle critique of the centrality of technology in modern life, most audiences and critics did not receive *American Graffiti* that way.[39] Some apprehended a "neon wasteland" where everyone, even the waitresses, exists on wheels and is "electronically linked through the pulsating beat of their radios," but most agreed with *New York Times* critic Aljean Harmetz that the film wistfully evoked the innocence of youth rather than the innocence of the United States as a nation.[40] Although Curt, who eventually decides to leave behind parochial small-town life for adult life in the 1960s, may have embodied Lucas's sense that "the fifties can't live" forever, the film is laced with optimism about the values of the 1950s that Lucas thought worthy of revival in the doldrums of the 1970s. As Lucas told *Film Quarterly*:

> I realized after *THX* [his first science fiction feature] that people don't care about how the country's being ruined. All that movie did was make people more pessimistic, more depressed, and less willing to get involved in trying to make the world better. So I decided that this time I would make a more optimistic film that makes people feel positive about their fellow human beings. . . . Maybe kids will walk out of the film and for a second they'll feel "We could really make something out of this country, or we could really make something out of our lives." It's all that hokey stuff about being a good neighbor, and the American spirit and all that crap. There *is* something in it.[41]

Central to Lucas's yearning for 1950s optimism was his cinematic aestheticization of that decade's car culture. Like a significant current in Western visual culture, which has long represented the human conquest of nature as "natural, right, then beautiful," the cinematography of *American Graffiti* softened and beautified US petroculture, rendering it an attractive and seductive element of the past (fig. 9).[42] A prominent film critic in the 1970s praised Lucas for poetically expressing "how the automobile has influenced the American way of life" and for aestheticizing the materiality of car culture through the film's effortless fluidity and its beautiful production design "inspired by the look of neon signs and the flow of auto headlights along city streets" in a way that was "pleasantly modulated, warming instead of punishing."[43] Another critic described the film as "a ghost-

Figure 9. George Lucas's dreamlike vision of US petroculture in *American Graffiti.*
Source: *American Graffiti,* Universal Studios, 1973; Universal Studios Blu-ray, 2012.

dancing, iridescent nightgown, a galaxy of pranks, games, thrills, and lights through which the gaudy cars weave and cruise like phantoms."[44] The *New York Times* called it a "dream landscape; the cars glide through the darkness in a strange, hallucinatory parade."[45] For its characters, the prospect of a dimmer, more mature future (visualized as a dull airport) renders their nights cruising in cars as "all the more an enchanted playground in which life is always a whirling carousel of delights."[46]

American Graffiti's blend of nostalgia and uncertainty about car culture struck a chord with American audiences during the energy crisis. The film was, and continues to be, wildly popular as an artful piece of Americana full of melancholy and romantic attachment to a certain middle-class experience of that culture.[47] It contributed to a wave of fifties nostalgia in the seventies that included *Grease* and *Happy Days.* After the turmoil of the 1960s and economic stagnation in the 1970s, many middle-class Americans romanticized what they perceived to be a morally simpler decade and the guiltless automobility that they remembered. *American Graffiti* became the third-highest grossing picture of 1973, with over $55 million in rentals and five academy award nominations, including best picture, best director, and best screenplay.[48] When Universal re-released the film in 1978, it added a further $63 million to its box office total. The film brought George Lucas to prominence and spawned a popular fifties nostalgia television show as well as a lesser sequel, *More American Graffiti,* in 1979. In 1995, the National Film Registry at the Library of Congress selected *American Graffiti* for historic preservation as a culturally significant motion picture.

Petrocultural nostalgia during the energy crisis appeared in a more aggressive and sinister guise in Twentieth Century–Fox's expensive, postapocalyptic science fiction epic, *Damnation Alley* (1977), which blended environmentalist concerns about earth's future with fantasies of a return to unfettered automobility and the patriarchal white family in a postapocalyptic future. Designed to be the studio's blockbuster film for 1977, *Damnation Alley* failed to impress either critics or audiences.[49] But as a studio product aiming to appeal to mainstream audiences, *Damnation Alley* is a window into the nostalgic cultural mood that the filmmakers hoped to exploit. Its plot revolves around a small group of soldiers who survive a nuclear war and must leave their damaged base in California to search for other survivors in Upstate New York, where post-nuclear life is apparently much less hostile. Fortunately, these men possess two special US military vehicles: amphibious tanks that are essentially big rigs with missiles. These machines provide protection from the hostile post-nuclear atmosphere and landscape. Over the course of the film, the group loses two soldiers but adds a young boy and a woman before arriving in post-nuclear Detroit. After a spectacular storm and a massive flood, presumably caused by the earth's damaged atmosphere, the group emerges from their ark in sunny upstate New York, where they are reunited with other survivors.

Damnation Alley imagined the American future as an idealized, middlebrow version of the past, including a regime of automobility, competition, and harmonious family life led by a patriarchal figure. It was nostalgic for a pre–energy crisis, pre–civil rights world where the mobile, white, middle-class family could live in peace and relative comfort, unbothered by the political demands of other groups or by shortages, ecological destruction, and social unrest. The film rather disturbingly imagines nuclear war as a cleansing event that sharpens competition for survival and success, and after which only the "best" elements of American society survive. The nuclear war thus enables white American society to reset itself to a fictive past, before the social and environmental upheavals of the 1960s and 1970s. Its survivors discover a strikingly idyllic post-nuclear existence that bears no apparent scars, only erasures.

As it progresses, *Damnation Alley* shifts its central dynamic from a multiracial group of surviving soldiers to a kind of white nuclear family, or what critic Judith Bloch called the "post-nuclear nuclear family."[50] Suspect or racialized characters, such as Sergeant Keegan (Paul Winfield), die off until only white Americans remain to form a surrogate nuclear family. George Peppard's character becomes the stern military father figure paired with Dominique Sandra's vaguely maternal "Janice," who takes on the role of surrogate wife and mother for the survivors.

Likewise, Jan-Michael Vincent's character evolves from young soldier into eldest son, and the boy (Billy) that the survivors find becomes his younger brother. Several scenes reinforce this family dynamic, including one in which young Billy is taught how to drive "the rig" by his surrogate father and brother, and the film's final sequence, in which the boys disobey their exasperated "parents" by speeding ahead into town on their motorcycle. *Damnation Alley*'s dramatic focus thus narrows to the family dynamic of these characters and the hope that they will survive as an intact family. In this way, the film posits the white nuclear family as the basic social unit of American society, echoing Wendy Brown's argument about the gendered dynamics of neoliberal selfhood that enable its male characters to thrive in an environment of harsh competition and extreme risk, none more so than a postapocalyptic land.[51]

In addition to fantasizing about the revival of the white nuclear family, *Damnation Alley* adopts a Social Darwinist vision of the future that asserts the centrality of competition to post-nuclear life. The film's opening text initiates this reading when it describes the ravages of the nuclear war on the planet and declares: "When this epoch began to wind down, the remnants of life once more ventured forth to commence the struggle for survival and dominance." The rest of the film charts this "struggle" in an inhospitable desert environment against mutated nature and against other human beings. As one critic pointed out, the demands of such Darwinian competition require the film's pioneers to suppress their humanity, explicitly refusing to regret past actions—including the nuclear war itself— and to mourn the death of their friends.[52] The only wholly human character, the African American Sergeant Keegan, bemoans the destruction of the earth, questions the utility of technocracy, and inquires about the gas mileage of the military vehicles. He is then dispatched by killer cockroaches, outlived by the more pragmatic pioneers who understand that life is a competitive struggle and adapt to thrive in it. These survivors find solace in their family unit, which enables them to keep moving, never stopping to think about the morality of their past, present, or future actions. Their only morality is survival, and their goal is to restore an idyllic past. The restoration of freedom as competition devoid of ecological, or other, restraints mirrors the valorization of the market in neoliberal thought from the period, as well as the valorization of cars as an expression of individual identity asserted against statist attempts to restrict energy use.[53]

Damnation Alley links its nostalgia for competition and mid-century middle-class security to automobility. It frames automobility as an essential element in the struggle to conquer nature and as central to the ideal American life for which

Figure 10. The automobility of US car culture is restored after nuclear war at the end of *Damnation Alley*. Source: *Damnation Alley*, Twentieth Century–Fox, 1977; Shout! Factory DVD, 2011.

the film longs. In effect, the film seeks to erase not only the social movements of the sixties and seventies but also the energy crisis and its potential consequences. It frames a nuclear holocaust as an unfortunate but effective solution to the energy crisis, as a fantasy for the restoration of unbridled automobility. There are no gas lines in this radically depopulated America, and the group keeps its vehicles running at a series of abandoned and apparently undamaged service stations along the route. At the end of the film, the boy is so fascinated by the wrecked automobiles in postapocalyptic Detroit (the heart of US car manufacturing) that he fails to notice an approaching storm. Despite one character's remarks about the cars as part of America's heritage—which he sardonically defines as "what you leave your offspring when you find out it doesn't work"—the film's final scenes of the surrogate brothers driving their motorcycle on the open roads of Upstate New York (fig. 10) negates any critique of US car culture in favor of a nostalgic celebration of its revival in a depopulated future. Nuclear destruction and depopulation have negated the anxieties and frustrations of energy scarcity in this new beginning. US car culture is reborn on the open roads of small-town America.

"Can't Drive 55": Trucksploitation and the Southern

The trucksploitation and southern film cycles combined the popular nostalgia of *American Graffiti* and *Damnation Alley* with populist resentments against elites accused of cheating hardworking Americans.[54] During the energy crisis, popular reportage often depicted protesting truckers as folk heroes—modern cowboys speaking up for the common man by resisting those in authority who threatened the freedom of the open road and the ability of working people to earn a decent living on it.[55] Indeed, truckers embodied what I am calling a petro-populist response to the energy crisis that framed it as a crisis manufactured by elites and demanded cheap fuel and open roads as the right of the people.[56] Trucksploitation and southern films reflected this petro-populist framing of the energy crisis. They used unrestrained automobility as a symbol and conduit for individual freedom and economic independence in an overregulated world, the birthright of every free American citizen. Many pitted their heroes—relatable, self-directed, fun-loving southern men—against state power, as symbolized by the corrupt, sadistic, or incompetent sheriff patrolling the highways and chasing the antihero at high speeds across various landscapes.

The decade's many trucker films and television shows tapped into popular fascination with what Shane Hamilton has called "the rural, hyper-masculine, neo-populist culture" of independent trucking.[57] The independent truckers of the cinema were the libertarian and anti-union heirs to the rugged individualism of the cowboy, whose truck (rather than horse) symbolized a mobile, autonomous, and masculine individuality. These films articulated a petro-populist neoliberal imaginary wherein state and union regulation of the (trucking) economy harms individual entrepreneurs and threatens the people's freedom of the road. As critic Richard Thompson noted in 1980, the truck functioned as a sort of family farm for the mid-century "redneck generation." It was a trucker's "home, his livelihood and place of business . . . his tool; and a fetish object he makes in his own image." The hero of the trucker film establishes his independence through his truck, the restriction of which becomes "the worst possible calamity."[58] It is a source of social identity and economic well-being, both of which are tightly linked. The truck was the perfect symbol of neoliberal entrepreneurial human capital in a populist frame.

Director Sam Peckinpah's trucker western, *Convoy* (1978), was the most popular film to exploit the populist trucker mythology of the 1970s. Based on a popular outlaw country song of the same name from 1975, *Convoy* tells the story of a band of defiant truckers who spontaneously form a convoy that becomes a social movement. The film opens on the desert roads of Arizona, where the audience meets the protagonist and eventual leader of the convoy, an independent trucker in aviator shades and a big black truck who goes by his CB handle "Rubber Duck" (Kris Kristofferson). After talking his way out of a ticket, Rubber Duck joins up with two truckers who will become his companions: a white ethnic nicknamed "Pig Pen" (Burt Young) and an African American known as "Spider Mike" (Franklyn Ajaye). As they speed along empty roads, the three are entrapped by the film's villain, "Dirty" Lyle Wallace (Ernest Borgnine), the sadistic sheriff who harasses truckers for pleasure. The truckers pay their fines but enrage Lyle by mocking him on their CBs. A racist, Lyle tries to arrest Spider Mike for vagrancy. The truckers resist, leading to a saloon-style brawl that forces the companions to flee to New Mexico, accompanied by a woman (Melissa, played by Ali MacGraw) and a few other truckers now implicated. So begins the ever-growing convoy. More truckers and a van of Jesus Freaks join the fray in New Mexico to form a miles-long convoy of populist dissent. Through CB chatter and superior driving, the truckers evade Lyle's efforts to thwart them. Soon, the governor of New Mexico (Seymour Cassel) tries to exploit the convoy's populist appeal for political gain. When his advisers dismiss the truckers as "trash," the governor declares, "This damn convoy has more public support than we do." Seeing though the charade, Rubber Duck and several others reject political co-optation to rescue Spider Mike, whom Lyle had beaten and imprisoned in Texas when he departed from the convoy to attend the birth of his first child. After liberating Spider Mike from prison, Rubber Duck sacrifices himself in a showdown with the National Guard, only to be resurrected at a funeral cynically staged by politicians in his honor. The convoy joyously returns to the road while Lyle laughs maniacally.

Convoy's populist vision of automobility in American life centered on the road, which the film imagines as a space of social belonging and play. All aspects of life in the film take place on or beside roads—the roadside truck stop, the cabin of the car or truck, the gravel shoulder, the gathering crowds on small-town roads, at a raceway. The rituals and realities of American life, of working, socializing, reproducing, worshiping, mourning, organizing, policing, and rebelling, take place in and around roads. This sense of social belonging and promise on the open road was not, of course, new. Walt Whitman's "Song of the Open Road" (1856) pictures the road as a pluralistic space of possibility and camaraderie.[59] *Convoy*'s vision is similar, with people from all walks of life, white and Black, male and female, joining together in a spontaneous movement on the open roads, where they form deep social bonds. But it is also decidedly populist, distinguishing between elites, who are represented as immobile, as passengers or as uncomfortable drivers who are willing, in the words of the film's governor, to "sacrifice the individual," and the people, symbolized by the truckers who journey together, stick together.

Convoy posits control of the road and the freedom to drive as a populist political struggle during the energy crisis, echoing the decade's trucker strikes and conservation politics. Although some scholars argue that *Convoy* and other trucker films sought to manage the threat of labor unrest posed by trucker strikes by featuring owner-operator heroes who disparage unions, the film acknowledges control of the road as a political struggle, but one in which individuals form a collective to defend the individual right to drive and earn a living unmolested by state power over mobility.[60] Political and police aggression in the struggle to control the road drive much of the film's narrative.[61] Each time the convoy adds new members, the group must overcome a new attempt by the law to halt the forward motion of the people.[62] In an echo of the discourse around trucker strike actions, the convoy's revolutionary potential is concretized in Rubber Duck's truck, which we learn is full of explosive chemicals when the federal authorities attempt a roadblock.[63] The authorities back off, believing that Rubber Duck is willing to die for his still-unknown cause, but they continue to try to control the roads, either by co-optation or force. They use all kinds of tools of the state, from surveillance networks, helicopters, and police cars to the military power of the National Guard, but they are always derailed by the revolutionary activity of the folky, fun-loving truckers. The film's use of military music reinforces this reading of a kind of populist war against illegitimate elite authority over the roads. Numerous and disparaging references to the 55 mph speed limit, which many Americans resented as an infringement on their freedom, broaden the film's antiregulatory appeal beyond trucker concerns to decry state attempts to control everyday drivers.

Convoy's populist defense of resistance to state control of the road leaned heavily on comparisons between truckers and cowboys, a common trope in 1970s popular culture. In her book *Trucker: A Portrait of the Last American Cowboy* (1975), artist Jane Stern compared the highway to "the long trail" and imagined truckers as the inheritors of the cowboy's "frontier individualism," whose "freedom to move knows no bounds."[64] That *Convoy* indulged in such comparisons is perhaps not surprising given Peckinpah's history with the genre. Early in the film, Spider Mike claims to have heard more stories about Rubber Duck than Jesse James, and in their first encounter Rubber Duck and Dirty Lyle lament the fact that independent men like themselves are a dying breed. These are classic themes in Peckinpah westerns, which he self-consciously updates for the age of roads and automobility to express his view that the "welfare state" had diminished "people's energies to create, to build, to the satisfaction of their own work."[65] *Convoy*'s truckers were the heirs to American cowboys, free to roam the open road. It claimed that the latter threatened the historic right of the people to individual freedom on the roads by attempting to fetter their automobility.

Despite its cowboy mythology, *Convoy* did not traffic in romantic elegies for the past as in *American Graffiti*. Instead, the film's mood leaned toward brash defiance of state authority in its assertion of the glory of fossil-fueled mobility. Peckinpah felt that "the trucks are as important as the actors" and spent hours filming them in order to lend the film visual and aural excitement.[66] As Marshall comments, "Some of the most beautiful and energetic footage centers on the trucks, their movement in particular—the shimmering emergence from heat waves . . . the lilting slow-motion dissolves in waltz tempo of vehicles and sand . . . the lyrical shots of 18-wheelers traveling in silhouette against a dying sunset."[67] One of the film's first images is of the exhaust from Rubber Duck's imposing black truck as it crests a hill and looms into full view. Toward the end of the film, trucks form an imposing line as they prepare to attack the prison that holds Spider Mike. The scene references western cinema, with the trucks taking the place of horses as expressions of the independence and power of the cowboy truckers. Trucks are both literally and figuratively agents of liberation, used to break through prison walls to free Spider Mike and to resist the National Guard's oppressive firepower in the film's climax.

Convoy locates the moral superiority of the truckers in their ability to harness the propulsive energies of automobility to live freely. Its fundamental conflict is between the constant motion of the convoy and the efforts of the sheriff and the federal authorities to control it. The film's visuals bask in the size and power of individual trucks and in the skill of the truckers to keep them moving. Their aptitude is a form of skilled craftsmanship that provides pride, agency, and income. Rubber Duck is often framed by the window of his rig, shot from a low angle, identifying him with the truck and lending him moral authority that the sheriff lacks (fig. 11). The sheriff's moral inferiority is marked by his inability to drive—to control the motion of his vehicle for his own ends. This visual rhetoric, alongside the absence of energy conservation as a legitimate social aim in an industry built on cheap petroleum, amounts to dismissing the energy crisis as the concern of effete elites inimical to the primal masculinity and freedom of movement that cheap fuel offers to independent truckers. For a film that rarely steps off the accelerator, fuel is remarkably absent, never used as it could have been as a dramatic threat to the existence and coherence of the convoy. The freedoms afforded by fuel, in the world of the film, are always primary, and state authority over them

Figure 11. Kris Kristofferson's "Rubber Duck" is an icon of anti-state, working-class cool in *Convoy*. Source: *Convoy*, EMI Films, 1978; Kino Lorber Studio Classics Blu-ray, 2015.

is always illegitimate. When the sheriff protests the attempted jailbreak of Spider Mike by saying that he represents the law, Rubber Duck replies, "Well piss on yah and piss on your law." The sheriff and his law are merely forces of oppression for the working-class men and women whose livelihoods rely on cheap fuel and free movement.

Just as it identifies the truckers with their rigs and fossil-fueled movement, *Convoy* frames automobility as the vehicle for the self-organization of their social movement. Over the course of the film, the size and diversity of the convoy grow into a mosaic of working-class discontent, or what the film's fictional governor calls "a hell of a cross-section of people." As the critic Richard Combs wrote, "The film and its ever-growing, road-hogging behemoth become a kind of wish ful-fillment machine, to which each individual and community it passes can attach their own banner of protest—a situation which opens up a surprisingly sunny vista of a land running riot with good-natured disrespect for authority, eager to be represented by such a freewheeling symbol but more likely to be exploited by the politicians who gather darkly in the second half of the film."[68] Key here is the propulsive energy of the convoy, which gathers energy, adherents, and the concerns of Americans as it moves through the country. The social and political strength and vitality of the convoy is rooted in its constant movement; to stop is to be threatened with imprisonment or co-optation.

Paradoxically, the film imagines the convoy as a manifestation of working-class consciousness while rejecting the kind of New Deal politics that such solidarity had historically entailed. It insists that the truckers' only shared aim is for freedom from regulation, symbolized by uninhibited automobility, driven by a feeling of disenfranchisement in a corrupt polity. When a journalist asks about the purpose of the convoy, the responses are varied. For Rubber Duck, "the purpose of the convoy is to keep moving." This could be read as hollow nihilism, or as a political statement about the threat that the energy crisis and regulation posed to trucking, the working class, and to American society in general. As other truckers in the convoy speak up about racial injustice, the 55 mph speed limit, big oil, and Watergate, it becomes apparent that this working-class solidarity is articulate but nevertheless built largely on populist grievance and alienation. Critics complained about thin plotting and character development, but the lack of a clear motive for the convoy was part of the point.[69] Aside from mimicking similar energy crisis protests, the individualized variability of the truckers' protest action suggests that working-class interests exist only at the level of individual trucker-entrepreneur, rather than at a larger class level. Hence, the film repeatedly emphasizes the fact that these are independent, nonunionized truckers who are fed up with their overregulated lives, and it lauds them for it.[70] What unites these truckers is the excitement and possibility that they find in fossil-fueled movement itself and want to preserve.

Through the pluralistic composition of the convoy, its articulation of contemporary political concerns, and its cynical view of political leadership, *Convoy* is an allegory of bicentennial America. It visualizes American democracy as a self-organizing, pluralistic convoy of jovial discontent. The truckers are bawdy but kind and intelligent, never posing a threat to the innocent. They are a fossil-fueled version of Whitman's ideal America, in which camaraderie and vitality exist and emerge through the force of propulsive energies on the road, in ceaseless motion. Throughout the film, vulnerability, weakness, and dissension are generally connected to life off the road, outside of the truck. When the convoy stops to hear the New Mexico governor's offer to take their concerns to Washington, the Woodstock-like atmosphere of the trucker gathering is threatened by the potential for political manipulation and division. Other moments of stillness are associated with danger and the derailment of the convoy by elites—being ticketed, a truck stop brawl with police, imprisonment, and the funeral for Rubber Duck. Automobility is always the vehicle for renewed freedom and vitality—a theme that likely resonated with Americans who sensed a threat to their automobility during the energy crisis as they waited in line for gas, spotted a new speed limit sign, or were cajoled to drive less.[71] As *New York Times* critic Vincent Canby commented about the popular car films of the 1970s, films like *Convoy* "show us that America is the first nation to find heaven on the highway. Heaven is no longer a destination. It's something experienced en route."[72]

In its setting and its recasting of a white working-class truck driver as the leader of a multiracial convoy, *Convoy* took part in a subgenre that emerged from the decline of the western genre amid the social upheavals of the late 1960s and 1970s: the "southern."[73] This new genre reflected key socioeconomic developments in the postwar United States, including the new economic power of the American South and West, the nation's urban crises, the civil rights and the women's movements, the waning of the counterculture, and the energy crisis. According to James Monaco, films in the southern genre "celebrate small-town self-reliance, everyday working-class life, and the ongoing struggle with the Man."[74] Many of these films adapted the themes of the western, including city versus countryside and individualism versus conformity, to contemporary southern car culture. Derek Nystrom argues, moreover, that they created a new cultural figure, the "good ole boy," which recast the image of southern, white, working-class men from the backwoods brutes of John Boorman's *Deliverance* (1972) to the "harmless, fun-loving, and well-intentioned" hero represented by actors like Paul Newman and Burt Reynolds.[75] After the success of Nixon's "southern strategy" in 1968, the good ole boy is a cinematic representation of a key southern Democratic constituency in the process of joining the Republican base in 1980. Crucially, in the southern he is in large part defined by his automobility and his resistance to corrupt state authority. A good ole boy loves his car, knows how to drive it, and enjoys outdoing dimwitted sheriffs.

The Sunbelt's postwar economic resurgence depended on the private automobile and the oil industry, so the conservation ethic posed a significant threat to the region's growing wealth and national political influence.[76] Films in the southern genre thus articulated populist resistance to state efforts to curtail energy consumption, prefiguring the "wise use movement" of the early 1980s.[77] Good ole boy characters mocked environmentalists and "hippies" while driving supercharged American cars that mocked the conservation ethic. The automobile is always central to the plot and to the action of the southern, as if to vindicate its status in the New South. Like the horse in the western before it, the car functions as the hero's special mode of mobility that enables him to overcome state oppression by outwitting, humiliating, or killing the sheriff.[78]

Burt Reynolds's immensely popular take on the good ole boy exemplifies the way in which the southern valorized automobility as a source of freedom and autonomy for the fun-loving men who loved to drive powerful cars and trucks. His characters shared some of Clint Eastwood's trademark self-sufficiency, though they also formed important bonds with others.[79] But they remained stanch individualists driven by their own moral code and the pursuit of freedom. Beginning with *White Lightning* (1973), many of Reynolds's most successful films from this period featured entrepreneurial good ole boys who evade the law with the help of a fast car or its equivalent. These films included *Gator* (the 1976 sequel to *White Lightning*), as well as the hits *W. W. and the Dixie Dancekings* (1975), *Smokey and the Bandit* (1977) and *Smokey and the Bandit II* (1980), and *The Cannonball Run* (1981) and *The Cannonball Run II* (1984). His films *Hooper* (1978) and *Stroker Ace* (1981) also heavily featured automobility, though they were less concerned with state power. In all of them, Reynolds embraces economic and physical risk. For instance, *Smokey and the Bandit* revolves around the illegal transportation of beer across state lines, a job with a potentially high payoff that is shot through with physical risk as Bandit (Reynolds) evades the law both on and off the road. Although Bandit forms relational connections, he is driven by the promise of money to buy a new truck.

In *White Lightning*, which Gene Siskel credited with triggering a trend of "car n' calamity Southern action flicks," Reynolds plays Gator: a moonshiner in prison who sets out to avenge his brother's death.[80] Gator's line of work immediately connotes entrepreneurial subversion of state authority over the economic freedom of individuals. When Gator learns that his brother has been killed by the town's corrupt and racist Sheriff Connors (Ned Beatty), he offers to cooperate with the federal government to collect evidence against the sheriff in exchange for his freedom. The feds lend Gator a supercharged car for the job.

Throughout the film, Gator's autonomy and masculinity are visually tied to his acumen for driving fast and to the car itself. Several chase scenes reinforce this link between agency and the automobile, allowing the audience to revel in the roar of powerful engines burning through petrol. For much of its climax—a chase between Reynolds and Beatty's sadistic sheriff—the film establishes the sheriff's moral inferiority by placing him in the passenger seat of the car in pursuit. Visually, the chase consists of powerful cars swerving and sliding across a lush southern landscape and close-ups of the protagonists in their machines. Whereas Reynolds asserts his moral superiority and individuality by ably driving his own car, the sheriff relies on another officer to drive for him. When Connors finally does take the wheel, he is clearly less skilled than Reynolds—he sits too close to the steering wheel, drives awkwardly, and falls for a trick that sends him flying into a lake, where he drowns. Although the film ends with the cynical suggestion that police corruption will persist, its identification with Gator and his car suggests that driving, like the illegal production of moonshine, is a viable way to resist state authority and to assert the autonomy of the good ole boy. The film suggests that the only way to be a free man, in the words of one character, is to "just keep your eyes on the road and the foot on the pedal." The lighthearted optimism of such films contrasted starkly with driving as alienated wandering in counterculture films like *Five Easy Pieces*. Indeed, the good ole boy's commitment to speed and economic freedom resolves the white masculine alienation of the early seventies. He is imagined as fully integrated in his community, often through oblique references to his lack of racism. The skillful good ole boy is imagined as a truly independent and self-possessed man, unlike the brutal sheriff who arbitrarily wields state authority over individual desires for the open road.

The moral distinction that *White Lightning* draws between Gator and Sheriff Connors further suggests the southern film cycle as an attempted rejuvenation of the image of the white working-class men of the New South. The film presents Connors, whose name is a direct reference, as a cinematic version of Eugene "Bull" Connor, the racist Birmingham police chief who violently repressed the civil rights movement in the 1960s. Sheriff Connors worries about the influence of communism and declares that the federal government will "integrate our schools . . . [and] get all our coloreds to vote." Gator, on the other hand, shares Connors's disdain for federal authority, but he is a post-racist man of the New South, defined and liberated by his car and his ability to drive. In contrast to environmentalist efforts to demonize cars and driving, *White Lightning* posits driving as a form of individual liberation and rebellion against state corruption.

Racing across America

Beyond mocking the authority of the state to control masculine automobility, car films reflected popular anger and frustration through heightened automotive violence and illegal road racing. Declining box office returns suggested that moviegoers had begun to tire of road movies by 1973–74, in part because Hollywood had produced so many, but also because they had become less relevant to people lining up for gas, worrying about oil dependence, or toying with oil conspiracy theories. As critic Stephen Harvey reflected in 1980: "Even if familiarity hadn't bred ennui, how was the road to keep on connoting freewheeling escape from the mundane, with gas costing what it now did?"[81] This brief waning of the road movie's popularity opened up space to express anger and anxiety about automobiles and state efforts to curb the worst excesses of US energy use. Reduced speed limits and new emissions standards particularly galled car enthusiasts, who felt embattled by environmentalists and others who had begun to use the energy crisis to attack cars as a threat to sustainability and community.[82] Thus, a reactionary cinema of high-speed chases, extreme automotive violence, and illegal racing expressed popular resentment about the perceived threat that energy conservation efforts posed to car culture. Cars became tools of titillating violence and excitement used to create chaos and to lash out against their potential obsolescence. Yet the sheer number of crashes and explosions belied ambivalence about the car, an anxiety eventually resolved in celebratory illegal road race films.

Gritty cop films like *Bullitt* (1968), *The French Connection* (1971), and *The Seven-Ups* (1973) first capitalized on the opportunity afforded by a new Motion Picture Association of America (MPAA) ratings system and more mobile film technology to depict violence and bring realistic, high-speed chases to the big screen. The chases in these cop films inflicted some damage on cars and their surroundings, but the excitement was derived from the risk of a crash and the drama of the chase rather than directly from violence. Over time, the tone of car chase films and the character of the violence involving automobiles changed. Filmmakers provided more violent, spectacular, and frequent car chases and crashes in a spate of car chase films wherein violence and destruction were the point. Such films included the surprise hit *Dirty Mary Crazy Larry* (1974), *Gone in 60 Seconds* (1974), *Death Race 2000* (1975), *Bobbie Jo and the Outlaw* (1976), *Double Nickels* (1977), *The Driver* (1978), *Deathsport* (1978), *Mad Max* (1979), *The Blues Brothers* (1980), and *The Junkman* (1982). In these, the car is a tool of violence against the social body, which is imagined as faceless, incompetent, or worthy of the protagonists' antiheroic scorn. Car chases are frequent and violent, leaving destroyed cars and infrastructures in their wake.

Car crash films resisted conservation discourses by pushing the limits of automobiles and using them to assert the violent will of their drivers. The 1974 hit *Dirty Mary Crazy Larry*, for instance, is essentially an extended police chase that celebrates its nihilistic, antisocial protagonists, Larry and Deke. The men are aspiring NASCAR drivers who extort a grocery store of $150,000 by holding the manager's wife and daughter hostage. As they make their getaway, they are joined by Mary, Larry's one-night stand. The rest of the film is a car chase in which the skilled and single-minded drivers outpace and outmaneuver the police. With numerous crashes and a soundtrack consisting largely of revving engines, the film clearly sides with its protagonists' willingness to do anything to achieve their private goal: to race cars. The trio is ultimately killed by a train after an act of hubris, but after ninety minutes of thrilling driving, the film's criticism of Larry's recklessness is rather meager.

Larry and Deke's motives echo those of many car chase film protagonists from the period. These and similar characters comprised *Homo economicus* figures who eschew the public or common good in favor of competition and the realization of their private economic interests, using automobiles to assert their will. The public is often the backdrop, or the victim of the mayhem created by the car chases and crashes, but the consequences of the havoc that they exact is never explored. The audience is instead encouraged to root for Larry and Deke against the police, despite their willingness to threaten violence against innocent people to finance their racing dreams.[83] Their reckless driving is framed as a show of skill and daring, and much is made of the superiority of their private automobile over the outmoded cars of the public police force.

Another surprise hit film, *Gone in 60 Seconds* (1974), brought the car crash film to a new level of destruction and made millions of dollars over a two-year theatrical run. Like *Dirty Mary Crazy Larry*, its success derived from its obsession with the movement and destruction of automobiles, although it also attempted more working-class respectability in its plot about principled car thieves. The film featured numerous high-speed car chases, including a forty-minute car chase across five towns that left dozens of mangled cars and wrecked city infrastructure in its wake. Reviews and promotional material focused on the scale and complexity of the climatic chase scene, always boasting that the filmmakers destroyed ninety-three cars to shoot it. Reviewers lauded its action as "madcap [and] rollicking," and audiences responded positively, turning the film into a resounding financial success largely through word of mouth.[84] Critics reported theaters full of enthusiastic moviegoers cheering for the film's real star: Eleanor, a 1973 Ford Mustang.

It seems fitting that so many people responded to a film so evidently in love with but also anxious about automobiles and American car culture so soon after the first oil embargo. Made by the owner of a junkyard, H. B. Halicki, the film featured a particularly revealing slow-motion sequence where Eleanor launches off a smashed-up car to evade police cruisers. The only use of slow motion in the film, the sequence features many different shots of Eleanor flying above the road. As critic Julian Smith remarked: "It's beautiful, the way the poor tortured old car—so young and shiny at the start of the chase—breaks loose in a literal and metaphorical flight of fancy, leaping ten times her own length while the twenty klutzy pursuit cars slam to a dusty death against the barricade of wrecked cars."[85] Given the many wrecked cars in its wake, the film's interruption of the illusion of linear time to linger on the battered Ford Mustang suggests a fascination with the automobile itself, as well as a desire to see how much damage it can sustain. The skill of the working-class driver is always on display here, but less so than the wanton destruction of dozens of cars, often in gratuitous detail as they roll, crumple, and fly off the road. Films like *Gone in 60 Seconds*, and the waste inherent in their creation, embodied a nervous but insistent resistance to the conservation ethic. While the state called on Americans to drive slower and less often, filmmakers drew audiences with extravagant chases that asserted the excitement of speed, destruction, and defiance of authority.

Media coverage of *Gone in 60 Seconds* emphasized the film's populist appeal by focusing on the personal story of H. B. Halicki, the film's director, writer, producer, star, and primary stuntman, as a story of working-class mobility. Celebrated as a "junkman turned movie mogul," Halicki came from a working-class background, the second youngest of thirteen children, and worked his way from pumping gas to running a junkyard and, eventually, to writing, financing, shooting, and distributing his own hit film.[86] A savvy self-promoter, Halicki grasped the importance of his working-class persona to the film's success. Speaking in 1975 about plans for a future film, he said, "It'll be written by a nobody, directed by a nobody, distributed by a nobody and it'll star nobodies."[87] His films, that is, were made by everyday working people for the working classes rather than for the social elites. They offered petro-populist excitement for audiences who felt threatened by energy crisis and conservation ethic debates.

Perhaps more than any other film, Paul Bartel's cult hit *Death Race 2000* (1975) embodies the theme of the car as a tool of violence. *Death Race 2000* is set in a neofascist, dystopian future, where the government stages a deadly automotive blood sport to keep its citizens satiated.[88] Drivers in this annually sanctioned violence receive extra points for killing pedestrians with their weaponized automobiles; the more brutal they are, the better they do. This setup aimed to satirize America's addiction to violence and speed, featuring devoted fans who sacrifice themselves for their favorite driver and "Euthanasia Day," when the residents of retirement homes are wheeled out to be fodder for the racers. The film's plot also concerns an effort by an underground group to sabotage the race by killing most of the drivers and taking the most famous driver, "Frankenstein" (David Carradine), hostage in an effort to overthrow the president. After ample instances of auto violence, dirty tricks, and high-speed racing, Frankenstein wins the race, overthrows the president, and kills the commentator who objects to his decision to outlaw the race.

Although *Death Race 2000* attempted a muddled critique of violence in American society, its celebration of weaponized automobiles and its abhorrence of big government stand out. The scenes involving murder by automobile and high-speed racing are shot for maximal excitement rather than reflective critique. The film's hero, Frankenstein, lives by the motto "Winning is the only standard of excellence left" in a society that requires absolute fealty to the state—a view that echoes the neoliberal valorization of free market competition over planning and regulation. The film's final scenes undermine any social critique when Frankenstein impales the president with his car and then runs over a reporter who says that abolishing the race could hurt Frankenstein's popularity.

Death Race 2000 succeeded financially and has since become a cult classic, yet its popularity, as contemporaries suggested, owed less to its satire than to its gratuitous automotive violence.[89] Critic Tom Shales cited *Death Race 2000* as an example of "the ruthless exploitation picture at its most efficient and primal." Shales acknowledged the film's thin effort at satire, comparing it to Norman Jewison's somber *Rollerball* (1975), as well as its interest in "the persistence of American auto-eroticism, and on humanity's general love of bloodshed." Nevertheless, he argued that the sheer amount of "murder by automobile" exposed its design "as a spectacle of kinetic titillation."[90] The *New York Times* agreed, noting that the film became "what it seems to have mocked—a spectacle glorifying the car as an instrument of violence."[91] As silly and over the top as it was, *Death Race 2000* connected with frustrated audiences by satirizing the violent potential of cars while simultaneously using them to express popular rage against injunctions to slow down, conserve, and let the state regulate the road.

Death Race 2000 also inaugurated a mini-cycle of road race films that mocked state energy conservation measures, particularly the 55 mph speed limit, by celebrating illegal racing for sport and a sense of rebellion.[92] For instance, schlock producer Roger Corman and David Carradine followed up *Death Race 2000* with the more realistic illegal road race film *Cannonball!* (1976), inspired by the real Cannonball rally. That same year Warner Brothers released *The Gumball Rally*, which director Chuck Bail said tried to realistically represent America's "racing underground." It featured police scanners, CB radios, and radar detectors, all of which had become popular consumer items for drivers looking to avoid highway cops. To achieve a realistic look, Bail insisted on filming on real roads driving at real speeds of 100 mph and above, often with a police escort in real traffic. Bail even selected his lenses to maximize the sense of speed and danger.[93] The film's advertising and dialogue positioned it as a populist rejection of the 55 mph speed limit. In one exchange, two characters complain that the speed limit imposed to conserve gasoline is unsafe because it is boring. Both agree that people need speed to stay alert and engaged lest they "fall asleep from the monotony" of their "air-conditioned, stereo-ed, floating capsule." They also despise "the way the world is going" toward safety rather than speed and risk.[94] Remarkably, then, a film reveling in the existential pleasures of the ultimate consumer item in the postwar United States—the automobile—seeks to position itself as an anti-consumerist manifesto, railing against the tedium of a regulated society. Although critics dismissed *Gumball Rally* as "nothing but one long exhaust pipe" and the "biggest, the gaudiest, the noisiest, the most lavish and mindless" car lust film of 1976, its sentiments resonated with popular resentment of state regulations over energy, which helped take Burt Reynolds's cross-country road race film to financial success just a few years later.[95]

Burt Reynolds's 1981 film *The Cannonball Run* united the themes of petro-populism, automobility, risk, masculinity, and neoliberal selfhood that variously appeared in other seventies car films. It featured an illegal cross-country road modeled on the real Cannonball races that, as in reality, it framed as a form of resistance to the very idea of fuel conservation and speed limits. It dealt in speed, openly mocking the 55 mph speed limit and the environmentalists who wanted to impose limits on American drivers. As one critic said of the film, *The Cannonball Run* "is not something one watches in any conventional way. One clocks it, yet it's not an easy competition to clock . . . the highway is alive with the sound of a loud musical score, spectacular car crashes, pursuits, sudden breakdowns and jokes. . . . Some of it is ingenious, and all of it is breathless."[96] The opening credit sequence of *Cannonball Run* and its sequel feature a Lamborghini speeding past police cars on an open road. Both times, the cars are driven by attractive women who vandalize the 55 mph speed limit sign. In the first film they spray-paint a red *X* over the limit in an act of faux rebellion (fig. 12), and in the second they plaster a 155 mph sign over top. In both, the supercar then speeds away while the camera emphasizes the open, empty roads of the American landscape, as if it is meant to be traversed by fast cars. The countryside is imagined only as a roadway for the excitement of speed while the ecological impact of highways and cars and burning carbon is wholly erased. In the early 1970s, the road was a place for self-exploration and a symbol of generational ennui. By the early 1980s, it was a

Figure 12. The Cannonball Run mocks the 55 mph speed limit. Source: *The Cannonball Run*, 1981, the Golden Harvest / Twentieth Century–Fox; HBO Studios DVD, 2013.

supercharged playground for speed-addled drivers to rebel against the supposed buzz-killing injustice of speed limits.

Cannonball Run framed resistance to energy conservation through the lens of masculinity. The film's villain is Arthur J. Foyt, an emasculated environmentalist and bureaucrat famous for banning the "use of electric toothbrushes during peak load hours." Early in the film, Foyt declares his opposition to automobiles and spends his time trying to shut down the Cannonball race to save the environment. He also desires the tree-hugging photographer, Pamela Glover (Farrah Fawcett), but she is stolen from him when Reynolds kidnaps her to ride in his racing ambulance. Foist channels his repressed desires into controlling the actions of others, in contrast to the "loose," self-fashioned personality of Reynolds. Reynolds asserts his control over Fawcett's character as well, nicknaming her "Beauty." Later, Fawcett asks Reynolds: "Why all this speed and trouble and risk?" He responds by telling her that his father was a coal miner who died right before he could retire and enjoy the fruits of his forty-three years of sacrifice. "I made up my mind right then," Reynolds says, "that I was going to go for it. I mean everything I ever wanted . . . I was going to go for it right then." This exchange is rich with subtext. It denounces coal mining, symbol of the nineteenth-century energy regime, as dangerous and the sacrifices that it exacts as meaningless. Reynolds has instead chosen to pursue his own interests, which are largely sexual and economic, enabled by oil. He decided to "go for it" by kidnapping the woman that he wanted and by illegally racing for money. A regime of oil-powered automobility is crucial here, since the scene suggests that it is this regime that makes Reynolds's self-fashioning and entrepreneurial spirit possible. A regime of coal is a regime of collective sacrifice. A regime of oil is one of individualism and the assertion of masculine will over both women and nature.

Conclusion

A crisis and reformulation of the postwar energy imaginary of the United States played out in popular 1970s films, though many critics did not perceive it. They instead generally dismissed these films as a culturally vapid trend that replaced story and character with a surplus of senseless action and violence.[97] Roger Ebert described the summer of 1976 as "the long summer of Good Ol' Movies in which automobiles were introduced for the transparent purpose of having their gas tanks explode, incinerating girls with two first names."[98] Gene Siskel sounded similarly annoyed when he called *Dirty Mary Crazy Larry* a "stupid car chase picture" aimed at people who are "not . . . nice."[99] The situation had not improved much by 1980, when he complained: "Never have more cars been destroyed in more summer movies than in the summer of 1980."[100] For most American film critics, the obsession with car crashes and chases signified a lack of ideas and imagination or, at most, a nihilistic culture rooted in a decade of political disillusionment. Whatever the reason, it seemed like machines and movement had supplanted drama, character, and intelligent social commentary. Some young directors agreed. Robert Zemeckis resolved never again to make a film featuring an automobile after his frustrating experience directing the cult satire *Used Cars*.[101] Despite the dismissive disdain that critics directed at car films, these movies were popular and profitable, and so they kept getting financed.

Exasperated critics reacted to the explosion of car films by criticizing their audiences rather than reflecting on larger sociopolitical roots of their appeal. Gene Siskel characterized their target as the "drive-in and teen-age crowds," while the *Washington Post* derided moviegoers for being "bored, undiscriminating thrill-seekers."[102] Another critic described the Australian crossover hit *Mad Max* as "ugly and incoherent and aimed, probably accurately, at the most uncritical of moviegoers."[103] Only the uncultured consumers of lowbrow could stand to watch such films. But Vincent Canby of the *New York Times* more insightfully thought that car films revealed "more about the state of mind of a large part of our union" than the art films celebrated in the major urban centers, discerning in them a rightward shift in the culture. But he also could not resist disparaging the southern and midwestern audiences that enjoyed movies like *Smokey and the Bandit* (1977), characterizing them as viewers who "respond . . . to the nonstop action (which is often just movement), to the colorful, heightened vulgarity of the language and . . . feel most at home in the country movie's principle setting: the automobile."[104] They exemplified, in other words, the tastelessness of the uncultured masses, especially in red states.

The ongoing obsession with automobility and car culture that American movie critics disdained ought to be viewed in the context of energy crisis and energy conservation discourses. Car films rejected conservation and living within limits in favor of speed, violence, and the liberty of the automobile as a tool against state oppression. They articulated a neoliberal structure of feeling that embraced individualism and risk against the monotony of a planned, steady-state society and advanced a petro-populist view of energy conservation as inimical to the right of the people to the freedom of the road. As anti-conservation imaginaries, car films established a cultural context in which many Americans confronted energy policy and anxiety over energy futures. The sentiments mirrored those that were eventually mobilized by the "bravura" of the Reagan administration in its rejection of conservation regulation and its militarized hypermasculine reassertion of US military and economic power in the 1980s.[105]

Car films suggest an uneven and ambivalent reception for the conservation ethic. Many Americans and their leaders embraced energy conservation, or the formation of some kind of conserving society, as a necessary way forward, while others resisted it as an illegitimate incursion on their freedom. Many car films articulated similar reservations. The resistance to conservation that they exemplified suggests a starting point for understanding the shift in energy conservation discourse from community to individual, from ecology to market.

Chapter Five

"Markets Born of Shocks"

NYMEX Oil Futures, Financialization, and Neoliberal Narratives

> The energy futures market, specifically the new crude oil futures contract,
> is still in its infancy but has great potential as part of the economic
> revolution.
> —DAVID S. LEWIS

The New York Mercantile Exchange (NYMEX) began its institutional life as the Butter and Cheese Exchange of New York on September 10, 1873. The exchange served as a hub for buyers and sellers to inspect agricultural products on-site to negotiate prices and exchange cash for goods. In June 1882, after a decade of financial struggle, expansion, and name changes, the exchange was renamed the New York Mercantile Exchange. Over the years, the NYMEX evolved to include new forms of trading and several charitable endeavors, including providing a form of life insurance for members and selling US Liberty Loan Bonds. The exchange moved into potato futures after the Second World War, though futures trading remained a minority activity at the exchange until the 1960s.[1] Throughout this time, it occupied a marginal position in America's financial landscape, overshadowed by larger institutions like the New York Stock Exchange, the Chicago Stock Exchange, and the Chicago Mercantile Exchange.[2] That would all change, however, when the struggling NYMEX pioneered energy futures trading in the 1970s and 1980s that reshaped the economy of oil and catapulted the exchange to elite status. Capitalizing on the energy crisis and a tidal wave of financialization, NYMEX's efforts exemplified the link between the energy crisis and the triumph of a neoliberal socioeconomic paradigm that claimed to obviate the need for planning through the power of free market choice.

The narrative justifications for oil futures on the NYMEX—the subject of this chapter—marked a key moment in the mobilization of energy and the energy crisis to empower a neoliberal view of markets. By disavowing regulation and planning as elitist, antidemocratic, and ineffective, these narratives echoed a populist notion of market participation emerging in public discourse that aimed to resolve the contradiction between public distress over rising energy prices and elite endorsements of market pricing to discipline energy consumers. For instance, when Ronald Reagan decontrolled the price of oil as one of his first acts as president of the United States, he claimed that price controls had created price instability and economic insecurity by fostering dependence on OPEC.[3] In a subtle jab at the Club of Rome's computer simulations and the forms of forecasting behind limits to growth and alternative energy discourse in the 1970s, Reagan's official energy policy document, *Securing America's Energy Future*, insisted that free markets were the only way to "chart . . . the Nation's energy path."[4] Markets, it said, were "a continuing plebiscite" conducted by "the American people themselves" that offered the flexibility to respond rationally and efficiently to changing conditions.[5] By participating in markets, Americans participated in the creation of energy policy and helped to shape the nation's future.

The rhetoric around oil decontrol, which the administration framed as a step toward "an energy policy focused on market realities," displaced both the emphasis on stability in postwar economic policy and the scarcity paradigm of 1970s energy crisis debates in favor of a neoliberal approach.[6] Rather than aiming for more or less energy consumption, the administration wanted American "citizens to have *enough* energy" and to leave it "up to them to decide how much energy that is, and in what form and manner it will reach them." By liberating millions of individuals to decide for themselves, free markets would "produce the proper balance of supply and demand."[7] Reaganomic energy policy imagined the energy crisis not as a crisis of scarcity or energy transitions but as one of economic productivity and dependence rooted in the state's misguided efforts to realize specific and stable futures.[8] In its valorization of the power of markets to solve the energy crisis and its embrace of risk, flexibility, and preemption, the Reagan administration paved the way for the establishment of crude oil futures contracts in 1983, a free market technology steeped in a financial understanding of the crisis designed to solve it through profitable speculation.[9]

This chapter demonstrates the consolidation of a neoliberal socioeconomic paradigm in relation to the energy crisis by tracing the origins of oil futures trading in the United States, as well as the legitimizing narratives that accompanied it. US financial exchanges first devised futures contracts in oil products in the immediate wake of the 1973–74 energy crisis, though they did not succeed until the late 1970s and early 1980s. Their eventual success, however, transformed the global oil and finance industries and helped to strengthen the broader and ongoing financialization of the US economy.[10] Futures traders and their advocates often claimed that oil futures had helped to end the energy crisis and bolster US geopolitical interests by helping to lower the price of oil to sustain the nation's "fossil economy."[11] The energy crisis thus provided an effective point of focus for advocates of financialization and lent their narratives symbolic heft. By helping to "solve" the energy crisis, they claimed, oil futures demonstrated the potential of neoliberal financialization to restore prosperity after the doubts about growth, productivity, and profits that haunted the 1970s.

Much of the analysis in this chapter concerns "neoliberal narratives," which were the stories, explanations, framings, and justifications that circulated to establish the necessity and utility of oil futures trading as a market-based response to the energy crisis. In light of the deep suspicion that many Americans have historically harbored toward futures trading, the exchanges, economists, and pundits in favor of oil futures expended an enormous amount of energy explaining themselves in the popular and professional press and elsewhere. Their narratives about the superiority of market mechanisms dovetailed with the larger narrative shift that Per Hansen has discerned in the decline of the Keynesian social state and the rise of neoliberalism. Whereas the grand narrative of Keynesianism imagined the capitalist market economy as an inherently unstable phenomenon in need of state intervention, the grand narrative of neoliberalism claimed that state intervention distorted the market and derailed its "natural" tendency toward efficiency and equilibrium.[12] The narratives about oil futures that are the subject of this chapter framed, aligned with, and advanced this larger shift to neoliberal financialization.[13] The institutional and discursive development of oil futures trading thus allows us to grasp more firmly the larger process by which the market came to be the "dominant social metaphor" of the late twentieth century.[14] Rather than supporting a conservation ethos and an energy transition, as many environmentalists and others had hoped in the 1970s, the articulation of a crisis drove the ascendency of finance and a neoliberal conception of markets as populist, democratic spaces—much like the open road in car films—where individuals determine energy policy through their choices.

This chapter first contextualizes these developments by narrating the longer history of futures markets in the United States and their overlap with the processes of financialization that energy crisis discourse bolstered. It then explores the development of oil futures contracts in the 1970s and 1980s, before turning to an analysis of the legitimating narratives that accompanied their creation. As a whole, the chapter shows how public discourse about futures trading in general, and oil futures in particular—from traders, experts, and the media—helped to legitimize the financialization of US energy policy. The much-celebrated oil futures contracts bolstered the emergent power of financial thinking as the United States entered a neoliberal era.

A Brief History of US Futures Markets

A futures market is a centralized marketplace that facilitates the buying and selling of futures contracts. A futures contract is a legally binding commitment to receive or deliver a specific, standardized quantity of a commodity at a designated location and date in the future, at a prearranged price that is paid at the time of delivery.[15] In essence, participants agree to a transaction in the future at a price decided in the present that is based on how their economic needs relate to their expectations about future price movements. As financial expert Edward J. Swan puts it, all such contracts amount to the sale of a "promise" to buy and sell at a future date that enables participants to "hedge" against adverse price movements.[16] Speculators enable hedging by assuming the risks that hedgers seek to mitigate. Abstract speculation, however, drives futures markets. In 1985, only around 1 percent of NYMEX futures transactions ended in delivery of a physical commodity; and in the twenty-first century, the percentage has become even more "negligible," according to market analyst Dominick Chirichella.[17] Most futures contracts are either settled in cash or liquidated by purchasing offsetting contracts.[18] As a result, every futures agreement has a winner and a loser, though small investors who speculate lose more money, and more often, than anyone else, and the exchanges win by selling exchange seats and charging a fee for every trade.[19]

Oil futures trading developed within a longer history of US futures market expansion that began in the mid-nineteenth-century Midwest, where futures contracts emerged to manage agricultural surplus and seasonality. Midwestern grain farmers suffered a constant cycle of boom and bust in which grain flooded urban markets immediately after harvest and decimated prices, only to be followed by scarcities and large price increases. American merchants adopted "to arrive" contracts—first developed in eighteenth-century Europe for the purchase of commodities arriving by ship—to circumvent this boom-and-bust cycle.[20] Spurred on by technological developments and the speculative markets for pork bellies and oats needed by the Union Army, the Chicago Board of Trade (CBOT) formulated rules for futures contracts in 1865.[21] Speculators hoping to profit from price fluctuations by trading contracts participated in US futures markets from their inception, driving the proliferation of "corners" on the Chicago markets in the 1860s and 1870s.[22]

The first era of futures trading expansion continued throughout the major commercial centers of the United States in the late nineteenth and early twentieth centuries. The New York Gold Exchange developed gold futures contracts in the mid-1860s that were also plagued by cornered markets and price manipulation. The New York Cotton Exchange (NYCE) opened a cotton futures market in 1870; and the New York Butter and Cheese Exchange, the forerunner of the NYMEX, organized forward contracts in 1872. In 1874 the Chicago Produce Exchange, the predecessor to the Chicago Mercantile Exchange, formed to trade forward contracts for butter and eggs.[23] Coffee became a futures commodity in 1882, followed by sugar in 1916 and cocoa in 1925. Metals—copper, silver, zinc, and lead—started futures trading in the 1920s.[24] By the early twentieth century, prices in many commodities circulated without the actual commodity changing hands, creating opportunities for speculative profit and ruin.

As it became the "dominant mode of commodity exchange" (measured by volume) in the United States by 1890, futures trading faced fierce opposition from farmers and populist politicians who felt cheated by traders and other financial interests.[25] The United States responded by formulating intellectual and legal justifications for futures trading, most importantly in Oliver Wendell Holmes Jr.'s majority decision in the 1905 Supreme Court case *Board of Trade v. Christie.* Holmes declared futures trading to be legal because speculation on the exchanges bore some relation to physical trade in the *possibility* of delivery and the hedging of real economic risks.[26] Holmes's decision also entrenched the power of incorporated exchanges by outlawing the "bucket shops" that had emerged to facilitate unofficial betting on the price movements of official exchanges.[27]

Attempts to limit futures trading persisted, however, leading Congress to create the Grain Futures Administration (GFA) in 1922 to oversee futures markets.[28] It banned futures trading outside of federally recognized exchanges and vested the United States Department of Agriculture (USDA) with oversight through the GFA and the Grain Futures Commission. The GFA testified to the dominance of agricultural commodities in US futures markets at this time, a reality also reflected in the Commodity Exchange Act (CEA), the 1936 update of the Grain Futures Act. The CEA banned options trading and created a framework for federally mandated position limits for speculators. Shortly thereafter, the USDA formed the Commodity Exchange Commission (CEC) to administer the Commodity Exchange Act.[29] The CEA and CEC remained the foundation of US futures regulation until the early 1970s.

After years of limitation imposed by these New Deal era regulations and the economic stability ensured by the Bretton Woods agreement, US futures trading entered a second era of expansion in the increasingly volatile economic context of the early 1970s, setting the stage for the financialization of oil futures. The decline of domestic oil production in the early 1970s and the arrival of the energy crisis joined several events to jolt the US economy by igniting inflation and uncertainty as the United States became a net importer for the first time since the nineteenth century. The Bretton Woods system that had governed global economic affairs since 1944 dissolved in 1971 after Richard Nixon announced that the United States would no longer honor its pledge to exchange gold for US dollars held by foreign nations.[30] Pundits fretted, but the financial industry smelled opportunity. As journalist Bob Tamarkin noted, the new volatility "caught the scent of the boys at the [Chicago] Merc, who reacted to it like a pride of hungry lions on the prowl."[31] Leo Melamed, the head of the Chicago Mercantile Exchange, promptly paid Milton Friedman $5,000 to write a defense of financial futures for the post–Bretton Woods world. A few months later, Friedman declared the death of Bretton Woods and an opportunity for futures in a new age of exchange rate volatility.[32] The Chicago Merc had introduced futures contracts in seven currencies by 1972, and futures trading took off.[33] As noted by Melamed, futures trading volume increased 2,100 percent between 1971 and 1991, mostly in financial products. By 1991, agricultural products made up only 19 percent of the 325 million options and futures contracts traded on US futures exchanges.[34]

A new regulatory framework meant to address the multiplying economic crises of the 1970s further aided the expansion and intensification of US futures trading. Amid rising food prices, rumors of Soviet grain price manipulation, and the development of "naked options" trading on commodity futures not covered by the CEA, the once-effective CEA suddenly seemed unable to regulate futures trading early in the decade.[35] Congress responded by creating the independent Commodity Futures Trading Commission (CFTC) in 1974 to regulate the futures exchanges using the "least anticompetitive means" possible.[36] The new agency's existence outside of the Department of Agriculture signaled the declining importance of agricultural commodities in futures trading, as well as the growing influence of finance to the US economy amid a crisis of profitability. The agricultural economist Gary Seevers served as one of the CFTC's first commissioners, having previously worked in President Gerald Ford's Council of Economic Advisers, along-side Treasury Secretary William Simon, economist Herbert Stein, and others committed to the neoliberal revision of US economic life.[37]

The process of financialization also benefited from the changing global economic position of the United States. The OAPEC oil embargo confirmed growing fears of national decline linked to the war in Vietnam by pushing the country into its worst recession since the Great Depression, undermining its global financial hegemony through the transfer of wealth from West to East (the "petrodollars" puzzle) and narrowing the productivity gap with Japan and West Germany. Indeed, the issue of petrodollars sparked particularly deep anxieties about declining US hegemony and the threatening prospects either of newly enriched Middle Eastern states buying up American companies and institutions or of those states exacerbating the outflow of wealth by *not* "recycling" their wealth back through the United States.[38] The energy crisis and the petrodollar issue exposed the extent of US dependence on imported oil and the new power of OPEC states to dictate higher prices that impoverished the United States. If, as Timothy Mitchell argues, "the economy as a coherent structure" emerged in the postwar period to create a new social order, the economic crises tied to the energy crisis created an opportunity to reconfigure the US economy in the language and logic of finance.[39] Futures trading innovation emerged as an avenue for value generation right when the United States lost control over the price of oil, the world's most important raw material. Futures markets offered, as anthropologist Caitlin Zaloom notes, contracts able to circulate as abstractions that "create a new source of value apart from the material goods that lend their value to the contract."[40] As the energy crisis made oil's future newly uncertain, the exchanges turned their attention to the possibility of profiting from the circulation of millions of "paper barrels."

Creating Oil Futures

Oil prices up to the 1970s had been too stable to make futures trading attractive, particularly in the tough regulatory environment of the CEA. The financial exchanges in the United States thus first introduced oil futures contracts in 1974, wasting no time in trying to capitalize on the destabilizing effects of the oil embargo. As the president of the Petroleum Associates of the NYCE told the *Wall Street Journal* in 1974, "futures trading doesn't work in a stable market," but it could help to "flatten out the peaks and the rises" in the unstable market that the oil embargo had created.[41] Thus, before the end of 1974, the NYCE started a crude oil futures contract and the New York Mercantile Exchange began trading futures for heating oil (gas-oil) and Bunker C oil.[42] Both contracts required delivery at the Dutch port city of Rotterdam, which had the largest crude oil storage and shipping infrastructure in the world at the time. Advocates pitched this first generation of contracts as a solution to the problems that the energy crisis either created or intensified, including profitability under high oil prices, security in the face of energy scarcity, and enhancing competition in the oil industry.[43] On the first day of trading at the Cotton Exchange, however, only two lots changed hands, foreshadowing the eventual fate of this first generation of energy contracts.[44]

By the early 1980s, the standard explanation for the initial failure of energy futures trading in the 1970s pointed to oil industry skepticism and insufficient volatility. Exchange officials and economists claimed that traditionalist oilmen refused to participate because they doubted the utility of hedging their risks in the futures markets and distrusted the intentions of traders with no history in the oil business. These oil futures contracts also lacked volatility—another word for risk—sufficient to attract speculative interest.[45] The Nixon and Ford administrations had tinkered with but ultimately maintained oil price controls, particularly in the controversial Energy Policy and Conservation Act of 1975, which opted to phase out oil price controls much more slowly than the oil industry and the exchanges wanted.[46] The relative price stability that resulted, the argument went, prevented energy futures trading from thriving. Only ongoing supply crises or complete decontrol would foment enough price volatility to move oil companies to hedge and to make speculators salivate. Persistent criminal activity on the part of oil futures traders, including convictions for "rigging" and "manipulating" oil futures markets in the mid-1970s, also contributed to their initial failures.[47]

The NYCE let its crude oil contract go dormant, but a desperate NYMEX rewrote its contracts in hopes of reviving them. Sanctions by the CFTC and a 1976 default in its core potato futures contract had threatened the future of the NYMEX, which relied heavily on potato futures. Its new chairman, Michel Marks, needed to find new commodities to trade, and fast. Believing strongly that oil could work, Marks and the NYMEX enlisted energy economist Arnold E. Safer to help rewrite their contracts.[48] They changed the grade of heating oil to be traded and moved delivery from Rotterdam to New York City. Two new contracts began trading in 1978; one quickly failed, but good timing eventually turned the other—for No. 2 heating oil—into a "dramatic success story."[49] It coincided with the renewed volatility and uncertainty stirred up by the Iranian Revolution and subsequent oil crisis in 1978 and 1979, as well as the outbreak of the Iran-Iraq War in September 1980.[50] By 1981 open interest for NYMEX's No. 2 heating oil contract hit seventeen thousand, which was well on its way to becoming "an industry icon."[51] Emboldened, other exchanges followed suit: the International Petroleum Exchange of London and the CBOT developed petroleum futures contracts in gas-oil, gasoline, and crude oil between 1981 and 1983.[52]

These efforts to develop oil futures reflected the ongoing financialization of the US economy in the 1970s. Derivatives trading proliferated in this period, particularly in futures, marking a move from growth to risk as the locus of profit in the economy.[53] Exchanges introduced futures contracts for US Treasury bills and bonds, as well as "cash settlement" futures (1975), which did not involve physical commodities by design. The latter included Eurodollars, interest rates, and stock index futures by 1982. Such innovations departed radically from early twentieth-century justification of futures contracts—namely, that participants could be said to be "contemplating delivery."[54] In each case, advocates justified new kinds of derivatives as a way to achieve greater economic security in an increasingly uncertain and volatile global economy by managing risk, a line of argument that would also be applied to oil futures contracts.

In search of solutions to the trenchant economic impasses of the 1970s, the United States government actively supported the process of financialization by encouraging the expansion of futures trading.[55] It is well known that the Reagan administration favored economic management by free market, but the Ford and Carter administrations also viewed futures markets as effective tools for enhancing competitiveness and for reducing inflation. Both supported the decontrol of oil prices in principle, saying that higher prices would stimulate energy conservation and innovation, though both settled for phased-in decontrol in the face of congressional opposition and concerns about the short-term economic consequences of immediate decontrol. Ford's Council of Economic Advisers consisted of neoliberal economists and traders dedicated to free markets, including William Simon and Alan Greenspan, who believed strongly in such deregulation. The Carter administration, for its part, deregulated several industries and supported the development of futures markets for their "pro-competitive effect."[56] Moreover, it was Carter's Council on Wage and Price Stability that supported the CFTC's three-year pilot program to reintroduce options trading in the United States, which led to the eventual relegalization of options on futures for the first time since 1936.[57]

In this pro-financial context, NYMEX continued to expand its energy futures options amid economic volatility. In 1981—as the Reagan administration finally decontrolled oil prices and US oil consumption fell for a third year in a row—the exchange introduced a contract in gasoline futures. By 1982 an oil glut flooded global oil markets and prices began to decline.[58] Crucially, the glut was interpreted by experts *not* as a permanent return to lower prices but rather as confirmation of oil's volatile future. In response to this long-awaited volatility, NYMEX introduced a crude oil futures contract in 1983 that quickly became the most valuable commodity futures contract in the world. In just four years, the daily volume of NYMEX crude oil futures trading jumped from 1,470 to nearly 950,000 contracts.[59] NYMEX followed its success with a 1986 options contract on crude futures that became the most successful options contract ever introduced by a US exchange.[60] These decisions helped NYMEX, which had faced "extinction" only a decade earlier, to become the nation's third-largest exchange. Indeed, NYMEX's official history frames the move from potato futures, where 80 percent of its members traded in the 1970s, to currencies and especially energy futures as its saving grace.[61] By 1986, NYMEX was billing itself "the energy exchange."[62]

Throughout this process, agricultural products lost their share of futures trading volumes to financial and energy contracts. NYMEX's move away from potato futures and toward energy and strategic metals emblemized the general shift of US exchanges away from their historic reliance on agriculture and toward geopolitically strategic commodities subject to vastly different temporalities and variables, as well as financial instruments with few ties to nature or the world of material commodities.[63] The stakes of futures trading had never been higher, given, for instance, the preceding decade's intense debates about oil supplies and prices. Nevertheless, the profit potential inherent in energy and financial futures drove the exchanges and their advocates to formulate legitimizing narratives that would link the expansion of futures trading, a key ingredient in the larger process of financialization, to the health and future of the US economy and nation. Ultimately, these narratives won out in the struggle to interpret the meaning of the energy crisis and helped to institutionalize neoliberal solutions to it.

Neoliberal Narratives

In light of historic and enduring suspicions about stock and futures markets, the US exchanges and their advocates sought to justify their rapid expansion in the 1970s to industry, government, would-be traders, and the public at large. When a 1978 public survey conducted by the New York Stock Exchange found that risk aversion prevented many Americans from participating in the stock market, NYSE chairman William Batten called for public education to avoid becoming "a nation of economically timid souls."[64] Futures trading advocates, who often attributed opposition to their cause to ignorance, responded to public, government, and industry skepticism by explaining the socioeconomic value and legality of trading futures contracts in currencies, bonds, and barrels of oil.[65] Joseph M. Burns of the CFTC, writing for the American Enterprise Institute, hoped that public acceptance of futures markets would usher in a "new economic order [. . . that] would spell gains in the vitality and dynamism of our free enterprise system."[66] Thus, throughout the 1970s and 1980s, the exchanges, economists, and other advocates circulated legitimizing narratives about futures trading in general, and oil futures in particular, that I will analyze under three broad themes: reason, the primacy of price, and power. These narratives offer a window into the logic of neoliberal financialization as its moment of ascendency.

REASON

The first narrative theme surrounding futures markets in the 1970s and 1980s cast them as a natural and rational response to the risks inherent in economic activity—the latest development in a long history of economic progress. This theme relied on evolutionary narratives of successive degrees of complexity that located the origins of futures trading as far back in time as possible. As the standard college textbook and how-to guide for new futures traders noted, "the roots of futures trading are as deep as commerce itself" insofar as commerce has always included some "concept of futurity in contractual arrangements."[67] Others used this broadly shared characteristic to place the origin of futures markets in ancient India, Greece, and Rome, as well as in twelfth-century France and England.[68] The Chicago Board of Trade's popular *Commodity Trading Manual* insisted that the traders and merchants of the ancient and medieval worlds "faced the same price risk and the same need for timely and transparent market information that today's business must confront."[69] All such accounts figured speculative risk as a transhistorical phenomenon inherent in human economic activity. If ancient Greeks used a form of futures trading to hedge their economic risks, then why should modern Americans not do the same? What they ignored, however, was that their ancient and medieval examples of contractual futurity ended in the trade and delivery of physical goods, a connection to material reality that by the 1980s was becoming increasingly thin.

Advocates of oil futures contracts in particular framed them as a rational adaptation to the new risks and uncertainties in the global oil industry. According to this narrative, OPEC's decision to wield the "oil weapon" had inaugurated an era of permanent crisis that would have to be managed by financial markets, not by government planning or energy transition. MIT-trained economist Philip Verleger Jr., a frequently cited expert, called the adoption of crude oil futures a "natural response" to the transfer of control over oil resources from Western multinational corporations to the non-Western governments of producing nations.[70] While critics claimed that futures markets created instability by encouraging speculation, Verleger and others insisted that they were merely a response to volatility rooted in foreign control over oil prices.[71] For instance, oil companies, facing a potential crisis of profitability in the 1980s, needed new hedging tools to manage their unprecedented exposure to volatile prices in a de-integrated and competitive global economy. The problem of uncontrolled oil prices, which Nixon Treasury Secretary George P. Shultz had once feared would raise "literally unmanageable" issues, could now be managed by hedging risks with futures.[72] The chance to claw back some of those profits, moreover, could also help to address America's troubling balance of payments problems, which had been exacerbated by the outflow of petrodollars.[73]

Narratives of oil futures as rational and necessary adaptations leaned heavily on a framing of recent US economic history as one of decline from mid-century stability to permanent volatility. Echoing the ways in which the energy crisis altered perceptions of mid-century America as a "Golden Age" of capitalism discussed in chapter 2, oil futures narratives reimagined the world before 1973 as a place where oil prices and supplies had been reliably stable, obviating the need for futures markets. When Western oil companies controlled supplies, production, and pricing, the story went, the world had had a sufficiently steady and stable supply of inexpensive oil to power global economic growth. For instance, as global oil prices collapsed in 1986, a NYMEX trading manual aimed at refiners and businesses keen to hedge their price risk attributed the postwar economic boom to the fact that most oil "flowed through the integrated channels of the major oil companies."[74]

This was a story that dutifully ignored the long history of boom-and-bust volatility in the oil industry that had driven oilmen to demand price controls and other forms of regulation earlier in the century. Instead, it blamed OPEC for disrupting the tranquility of Western corporate control over oil and for generating prolonged instability by attempting to maintain higher prices amid surpluses in the 1980s.[75] The result, NYMEX claimed, was that "the only certainty in today's market is uncertainty . . . risk management is, therefore, the key to long-term performance."[76] A study on oil price forecasting for investors and policy makers coauthored by Cambridge Energy Research Associates and Arthur Andersen & Co.

agreed: "Uncertainty—and volatility and turbulence—are inherent in the future of oil." In this climate, firms and individual investors would have to embrace flexibility and adaptability in order to "turn uncertainty into opportunity."[77] By destroying the existing system of oil extraction, production, and circulation, OPEC stood accused of destroying oil price stability forever, rendering free market speculation necessary to cope with permanent crisis.

Despite their commiseration about the uncertainties of living in a state of permanent crisis, the US exchanges and their investors knew that oil futures trading needed volatility.[78] Price movements, they claimed, had to be frequent, multidirectional, and unpredictable to make hedging rational and speculation possible.[79] Crisis, that is, was beneficial and ought to be normalized, rather than fought and overcome. However, if futures markets profited from degrees of volatility and risk that many Americans abhorred and that even the exchanges felt compelled to justify, what public good could oil futures trading offer beyond enriching traders and their firms?

Legitimizing narratives for oil futures linked two purported benefits of futures markets to the larger discourse of rationality: hedging and price discovery.[80] The hedging function was the most widely advertised social benefit of futures trading in oil or any other commodity.[81] By allowing large producers and consumers of oil to purchase contracts for oil to be delivered in the future at a guaranteed price in the present, the futures exchanges allowed those businesses to plan their expenses and manage the risk posed by the possibility of deleterious price movements in a volatile commodity. An airline could chart its fuel costs months into the future without having to worry that prices would go up unexpectedly, though in doing so it risked missing out on a price decline. In the 1980s, NYMEX frequently advertised its contracts as hedging tools while blaming "market forces" for generating the risks that it now offered to help manage.[82] Indeed, given the dogma of inevitable volatility, NYMEX often opined that to choose *not* to hedge was a greater gamble than trading futures. By the early 1980s, the market forces that conservation advocates had once embraced as a way to chasten US consumption were now framed as unfortunate but unavoidable realities that could only be *managed* through further financialization.

To posit hedging as a beneficial service of futures markets also required a defense of speculation, given the inextricable links between the two: there must be a willing speculator to assume the risks that the hedger wants to mitigate. Given longstanding suspicions of stock and futures speculation in the United States, which popular memory linked to the Great Depression, advocates took special care to differentiate between gambling and the role of speculators. Whereas the former served no productive social purpose, speculators were purported to ensure the smooth functioning of markets.

The difference between gambling and "legitimate" speculation lay in the source of the risks that the speculator assumed and the place of reason in the speculative activity. As one popular investment guide asserted, legitimate speculation, unlike gambling, dealt with "risks that are necessarily present in the process of marketing goods and services" in capitalist societies.[83] In the case of oil, a refiner risked exposure to fluctuating oil prices. By agreeing to assume some of the refiner's price risk, speculators enhanced the liquidity of the oil market and enabled the market to perform its social function. Moreover, by providing liquidity, according to an American Enterprise Institute treatise, speculators "enhance[d] the efficiency of markets for future transactions."[84] This argument echoed the 1905 Supreme Court decision in favor of futures markets as a legitimate risk management institution. Gambling, in this framework, was an irrational activity inimical to futures markets' creation of "anticipatory prices that reliably guide the optimal allocation of resources to the production and consumption of commodities."[85] Exchanges and experts therefore valorized rational speculation based on real economic risks that they claimed to be a permanent feature of the future; market speculation emerged as the only reliable guide to the allocation of resources according to prices.[86]

The defense of speculation as rational and necessary also invoked price discovery, the second purported benefit of futures markets. Futures markets, according to a 1985 report from the Federal Reserve Board and the CFTC, were useful financial technologies because they "discovered prices" by providing incentives, namely risk avoidance for hedgers and profit for speculators, to gather information in a centralized marketplace. In order to define their position in the market, participants collect and assess information to predict how prices might move in the future.[87] The constant buying and selling of contracts at different prices constituted a process of information collection, processing, and dissemination that purported to guide the "allocation of real resources via production, inventory, or other decisions."[88] Thus, although NYMEX preferred to emphasize the more sober hedging function in advertisements aimed at industry players, economists praised futures speculation as a boon to "the informational content of prices."[89] By gathering information to form rational price expectations, they argued, futures markets enabled firms to plan production, thereby stabilizing fluctuations of production and consumption over time. In this framework, futures markets turned intuition (speculative expectations about an unknowable future) into fact (the price of oil). Price, then, was far more a product of anticipation than a reflection of present supply-and-demand factors, and speculation on futures markets was essential in "discovering" it.

The notion that futures markets "discovered" the real prices of commodities, and that those prices were inherently anticipatory, paralleled the ascendant neoliberal conception of markets as superior information processors that attracted so many supporters looking for a way out of the decade's economic impasses.[90] Neoliberal epistemology tends to emphasize the complexity and inscrutability of nature as an argument for the free market and against planning.[91] No single government, individual, or company can efficiently synthesize enough information to predict and plan for the future without the help of markets because the degree of complexity is too high. There are too many factors to be considered for such centralized planning to work.[92] Instead, neoliberalism imagines markets as self-organizing systems that thrive on chaos to produce value through the movement of prices in markets. The paths of those movements are the only reliable guide to planning for the future, which must constantly be readjusted in light of new information.

Energy crisis discourse was a key arena to advance this high-efficiency, information-processing conception of the market. The crisis bred a sense of urgency about managing the now-uncertain future by forecasting possible futures and planning potential responses to them. Would oil be abundant or scarce, cheap or expensive, benign or corrupting to social organization and values? New energy technologies, conservation measures, and hopes for a low-energy transition all depended on forecasts to motivate change in the present. Neoliberals, however, rejected calls for scenario-based planning by denying the very possibility of collective planning. Only the market, composed of millions of self-interested individuals, could manifest the future in the present by processing information. It was the only viable planning tool, given the chaotic and complex nature of reality and the inevitability of crises.

Milton Friedman, in many ways the public face of neoliberalism in the 1970s and 1980s, often criticized centralized projections and planning as distortions of market pricing. As he declared in 1978, "The crystal ball is inevitably cloudy, that . . . is the nature of the world we live in." Predictions about the energy future could never be right. What US energy policy needed instead was "an adjustment mechanism that will enable us to adapt to what happens as it develops": the price mechanism of free markets.[93] Friedman argued that accurate prices emerge only as millions of people make "their best guesses about the future," which are continually adjusted as the market reacts. He insisted that the information content of prices was far more reliable than reserve estimates, which he claimed to "have no confidence in."[94] OPEC's greatest sin, according to Friedman, had been to artificially inflate prices by fiat. The sin of the US government in the 1970s, he thought, had been relying on reserve estimates and projections based on unknowable dynamics of supply and demand. Oil prices had to return to free market dynamics, not the political aims of OPEC or the price and conservation targets of the US government. In 1983, Friedman predicted in *Newsweek* that free market pricing would end the energy crisis and reveal a lower price for oil.[95] His assumption that scarcity is an economic rather than a geological and biological concept became the dominant lens through which energy was understood as oil futures trading took off.

Exchanges and economists in the United States posited futures markets as the most rational response to a volatile era in the global economy, which enabled the exercise of reason in hedging and price discovery, despite all appearances. Economists insisted that they came closer to "perfect competition" than any other capitalist institution and so could be trusted to reveal the true value of commodities by compiling and disseminating all information relevant to its price. As one study noted, "There are many buyers and sellers of a homogenous commodity in a buzzing hive of information, their transactions essentially unregulated except to ensure that no single participant amasses enough market power to manipulate prices."[96] According to another economist, futures markets "closely approximate the conditions necessary for perfect competition" insofar as their buyers and sellers conduct "arms-length" transactions of a homogenous commodity in "a well-regulated arena."[97] By flattening and standardizing as many elements as possible—delivery location, contract lengths, and commodity grades—the futures market created a space of pure expectation in which no allegedly extraneous material or logistical factor could shape the declaration of price other than expectation based on a specified set of dynamics. As I will discuss, the notion that futures markets facilitated impersonal, competitive transactions drew its rhetorical strength from markets' opposition to OPEC, which was taken to be the ultimate example of secretive and anticompetitive pricing, that is, of transactions conducted for political gain. By continuously broadcasting prices, futures markets were purported to enhance competition by allowing all market participants to be as informed as possible to make rational decisions.[98]

The Primacy of Price

Support for futures markets as instruments of price discovery dovetailed with an emerging set of narratives about the energy crisis and oil futures markets that posited the primacy of price in energy policy. Advanced mostly by economists who dismissed environmentalist arguments for planning to avert looming limits to growth, these narratives posited that price mattered more than material scarcity in understanding why the energy crisis had happened and how the United States could move beyond it. They reframed the energy crisis as a financial crisis that could be overcome by the right financial policies and practices.

The dominant narrative of the energy crisis in the early 1970s said that it was a crisis of geological scarcity caused by the rapacious consumption and wastefulness inherent in US capitalism. Gasoline shortages and skyrocketing prices signaled the fact that the pursuit of mass production and consumption in the twentieth century had begun to surpass the resource limits of the earth. For more than twenty years, petroleum geologist M. King Hubbert had been predicting peak oil production, and in 1972 the Club of Rome's sensational *Limits to Growth* report predicted devastating resource scarcities.[99] With these framing discourses in mind, media reports and expert commentary chastised the United States for its "astonishingly profligate" use of the earth's resources. As the *New York Times* declared in December 1973, the message of environmentalists was being "carried by a powerful medium: soaring energy prices."[100] This narrative called for fundamental changes in the way Americans related to the earth through energy transition, planning, and the development of a society committed to the conservation ethic, though some prominent voices saw the free market as a way to achieve these goals, as explored in chapter 3.

The neoliberal financial paradigm for the energy crisis differed sharply. Its advocates argued that energy prices confirmed the message of economists, not environmentalists. Rather than signaling absolute scarcities, high oil prices were the consequence of failing to rely on market pricing to allocate supplies and stimulate new production.[101] Existing scarcities were unnecessary, the result of domestic prices that had been kept artificially low, encouraging overconsumption, underproduction, and dependence on imported oil. The federal government had failed to handle what was essentially "an acute problem of financial management."[102] Had US oil production been subject to free markets rather than price controls, the argument went, there would not have been a crisis because scarcity cannot exist in a free market where prices rise to eliminate it and fall in response to surplus.[103] This was not a new argument, but it gained influence as a justification for decontrolling oil and for using futures markets to shape oil prices at the same time that conservation nationalists and their allies were arguing for greater reliance on free market pricing to discipline US energy consumption.[104] For the financial paradigm, like the conservation nationalists, the "crisis" in the energy crisis had not been exceeding the limits of the earth's resources; it had been ignoring the laws of the markets, which eventually resulted in losing control over the price of oil to OPEC as imports continued to rise.[105] It implied, as will be discussed, that managing future crises required reliance on markets to bring influence over prices back to the United States.

As the decade wore on, the shadow of scarcity also gave way to the problem of price in debates about the effect of the energy crisis on the geopolitical power of the United States. The high price of oil had certainly been a factor from the start in concerns about inflation, recycling petrodollars, and a defeatist national mood, but it was often shadowed by the fear that oil supplies could be cut off or that reserves could continue to decline in accordance with the limits to growth, as in a 1979 CIA report concerned with "the limited nature of world oil resources."[106] Over time, however, government energy experts began to worry more about the effect of high prices on US power than absolute scarcities, while some analysts warned that persistently high oil prices could damage key Cold War alliances with other oil-consuming nations.[107] One congressional report considered an energy futures market to be an effective tool to provide short-term price stability in case of minor supply disruptions, though the government would have to look elsewhere to overcome dependence on imported oil.[108]

The narrative of price gained strength in the early 1980s as oil gluts buried popular fears of scarcity. In February 1983 the state-owned oil companies of Britain and Mexico cut their posted prices below OPEC's price, facilitating the oil price decline that had begun in 1981 and continued to a final plunge in 1986. As the glut swelled, economic recovery seemed to be on the horizon, and it became much easier to dismiss the previous decade's concern with scarcity, conservation, and alternative energy and to argue instead that prices should fall as surpluses flooded markets.[109] *Business Week* dismissed concerns that recovery meant a return to high consumption when it proclaimed: "Conservation has been institutionalized: people expect to drive their small, fuel-efficient cars for a long time. . . . Home insulation is not going to be ripped out once it is installed."[110] Environmentalists and policy makers, it implied, should not fear a return to profligate consumption as oil prices fell. As crude oil futures began trading on NYMEX in March 1983, CBOT economist David J. Hirschfeld said that they heralded the potential "birth of a revolution in the petroleum industry, akin to that which has been witnessed in the financial community."[111] Financialization, it was hoped, would remake the oil economy in the interests of a US economy and state, presumed to need plentiful *and* cheap oil.

The emerging primacy of price provided narrative support for new oil futures markets, which their advocates claimed could solve the energy crisis through their effect on prices. A key moment in this narrative occurred in the 1978 hearings of the Energy Subcommittee of the Congressional Joint Economic Committee, which explored the question of the world's future oil supply from an intentionally "optimistic" perspective. In his opening statement, the chairman of the Subcommittee on Energy, Democratic senator Edward Kennedy, noted that conservation remained important, but that "the greatest problem is rapidly rising prices," which caused inflation, hurt the US dollar, and prolonged the global recession.[112] The testimonies of energy economist Peter Odell and astrophysicist Thomas Gold cast doubt on the reality of oil scarcity, while economist and NYMEX consultant Arnold E. Safer argued for privileging oil prices over politics.[113] Safer argued in a book published the next year (with an endorsement from Kennedy) that the scarcity paradigm acquiesced to OPEC's higher prices by rendering them inevitable. Indeed, many who supported an energy transition did applaud higher oil prices as incentive to conserve oil and develop alternative energies. Safer insisted, however, that "the scarcity thesis" was a self-fulfilling prophecy; it was "like a receding horizon: no matter how rapidly you move toward it, it is still the same distance away."[114] That is, accepting scarcity in the present only led to more in the future.

Safer advocated the establishment of an oil futures market to make the oil-pricing process more competitive as a way to undermine OPEC.[115] Lower prices, achieved through US markets, would boost the confidence of its main trading partners and allies, who could not rely on US economic stability as long as OPEC decided how much oil was to cost.[116] Rather than OPEC's "mercantilist conception of oil pricing," where prices are negotiated based on perceptions of the fair price that are filtered through political interests, the neutral competition of the free market would decide.[117] Jimmy Carter's emphasis on limits and conservation merely capitulated to OPEC, whereas futures markets could provide a way to reassert the waning economic power and global leadership of the United States. Safer concluded his book on oil policy: "We have been confused and fearful for five years; it is time for bolder action."[118] Such boldness pervaded coverage of

NYMEX crude oil futures trading in 1983, which clearly revealed the geopolitical and economic roles for futures trading: to rejuvenate US markets and undercut OPEC, the much-maligned threat to US economic freedom that had plagued the national consciousness for more than a decade.

POWER

The third legitimizing narrative for oil futures markets revolved around the restoration of US geopolitical power. The energy crisis had threatened US hegemony by worsening inflation and causing the unprecedented outflow of billions of US dollars to OPEC nations. The restoration of US hegemony, analysts understood, depended on "recycling" these petrodollars back into the US economy to mitigate the balance of payments crisis that they had created, though they fiercely debated how that process would work.[119] It also depended on lowering oil prices to stem the tide at the source, which meant finding a way to undercut OPEC's efforts to maintain higher posted prices.

After the establishment of crude oil futures contracts on the NYMEX in 1983, analysts praised oil futures as a finance weapon poised to disrupt OPEC's price setting power, which it had sought to use to achieve its own geopolitical ends.[120] I use the term "finance weapon" to mirror the language of an "oil weapon" that the US government, analysts, and media used to describe OPEC's actions in the 1970s. It expresses the hegemonic dynamic of financialization and oil futures as attempts to eradicate the threat that the energy crisis posed to the US economic empire. Financial institutions, traders, and the media imagined oil futures markets as a way to take pricing power back from OPEC, which they narrated in terms of futures trading as being a more democratic and fair form of price discovery than price setting. Crucially, this narrative *assumed* that the futures market price would be lower than OPEC's price—an assumption rooted in an imaginary of abundance and faith in the rationality of speculation to recognize it. Rather than using crisis to reduce dependence on foreign oil through conservation and alternative energies, the finance weapon mobilized the energy crisis to argue for the reassertion of US geopolitical dominance through the financial technology of futures markets.

Analysts worried constantly about the implications of US dependence on OPEC oil after the embargo demonstrated the cartel's power to use oil supplies and prices to achieve political goals.[121] They knew that oil gluts could hurt OPEC's ability to set prices, but how to capitalize on global oil surpluses was the domain of the financial exchanges.[122] The exchanges and the financial industry, for their part, framed futures trading as a way to undercut OPEC by stressing their more open and democratic nature. If, as the *Oil and Gas Journal* claimed in 1981, "futures prices are an open means of ascertaining free market values," then the creation of oil futures contracts would bring market prices to oil, which had always been decided by vertically integrated oil companies or biased "oil sheiks."[123] When NYMEX's crude oil contract took off, commentators openly speculated that this new form of transparent and rational pricing threatened OPEC's much-loathed reign. *American Banker* called crude oil futures "the ultimate threat to OPEC" because they were offering free market "public pricing" rather than allowing a cartel to charge "whatever they could get" for their oil.[124] According to one futures trader, the open outcry system, by which traders negotiated deals verbally on the floor of the exchange, "is the prime example of a free market where the sale goes to the highest bidder, and the product moves from the lowest seller," suggesting a sense of rationality and neutrality to market prices that ignored the wants and whims of powerful cartels.[125]

Even so, how were futures markets supposed to redirect power over oil pricing away from OPEC? In the late 1970s, Saudi Arabian light crude served as the benchmark for crude oil prices, having dethroned Texas crude in the 1960s. This meant that OPEC's posted price informed the price for oil sold on contract, which represented the majority of oil sales until the 1980s, and on the spot markets, which referred to the posted price for its transactions.[126] But NYMEX used American crude oil, West Texas Intermediate (WTI), as the basis for its crude futures contract. If its futures contract could become sufficiently influential, there was a good chance that the higher-quality WTI crude could supplant Saudi crude as the pricing benchmark, marking the return of Texas oil to world benchmark status.

Drawing on the claim that futures markets "discovered" prices, NYMEX worked to achieve benchmark status by touting the discovery of the "real" price of crude on its exchange floor. The exchange and its supporters claimed that the futures market transparently broadcasted the price for WTI, which meant that it offered the real price of oil to the public for the first time, whereas it had previously been decided privately by Western oil companies or, more recently, OPEC nations. The attraction of the real price, they expected, would turn WTI futures prices into the benchmark for sales on the spot market, where increasing amounts of oil purchasing occurred, and where the United States bought 25 percent of its foreign oil by 1984. For those who bought oil by negotiated contract, including Western oil companies, a lower futures price could serve as a reference price in negotiations, again to undermine OPEC's ability to dictate.[127] In the weeks leading up to

NYMEX crude oil trading, OPEC reduced its posted price from $34/barrel to $29/barrel for the first time, foreshadowing its crumbling pricing power.[128] Near the one-year anniversary of NYMEX crude oil futures, one oil company trader remarked that the contract either "puts steam into the [spot] market or it pulls the plug on the market."[129] By the early 1990s, oil analyst and historian Daniel Yergin could claim, "With the rapid rise of oil futures, the price of WTI joined the gold price, interest rates, and the Dow Jones Industrial Average among the most vital and carefully monitored measures of the daily beat of the world economy."[130]

The dominant media narrative about crude futures trading in the 1980s echoed NYMEX rhetoric by drawing a contrast between what it considered to be the transparent and democratic process of price discovery on the futures markets and OPEC's secretive and allegedly irrational posted prices. NYMEX chairman Michel Marks announced that futures markets let "people know that the price is being determined by many people instead of 13 oil ministers sitting around a table, cloaked in secrecy," while another trader claimed as early as 1984 that the NYMEX price had "become the new benchmark for oil. . . . A visible and centralized pricing mechanism has been put in living rooms and offices around the world."[131] The *New York Times*, similarly, praised "the bellowing on the floor of the New York Merc" for shaping prices more than the nationalist "frame of mind of Arab sheiks."[132] Open outcry made the price of oil transparent and available to all, even in spite of lurking concerns about the irrationality of traders and the chaos that many people perceived on the exchange floor. As *Newsweek* put it, oil futures clarified "oil price mechanics" by taking them "from behind closed doors and into the open pit of a free exchange."[133] These narratives framed futures markets as spaces of democracy and freedom in contrast to the anti-democratic structure and secrecy of the cartel. To financialize oil futures was to democratize oil pricing.

In this way, oil futures narratives in favor of financialization echoed the petro-populism of anti-conservation car cinema, suggesting a relation between petro-populism and neoliberal economic thinking. In place of fast driving as an assertion of the people's freedom to roam America's roads unbothered by the state, the financial press presented the futures markets themselves as democratic spaces that enable the people, represented by futures traders, to resist the tyranny of unaccountable bureaucracies by pursuing their own interests. Throughout the 1980s, newspaper accounts dutifully soothed popular anxieties about the trading pit as a den of thieves by framing their apparent chaos as actually "controlled" by the traders' desire for speculative profit or hedging risks. These financial motives ensured the exercise of rules and reason beneath the veneer of anarchy.[134] Accompanying visual representations of the pit often focused on a lone individual against the backdrop of the crowd. Often echoing the visual structure of Norman Rockwell's famous painting *Freedom of Speech* (1943), in which the town hall is a space of political freedom, media photographs visualized the futures exchange as a space of economic freedom. The trader pictured prominently in the foreground of these images speaks while everyone around him is typically quiet, as in Rockwell's image. Within the chaos of the market, these images suggest, there are individuals freely buying and selling at prices freely and openly agreed on in the dynamic markets of America, the town halls of the 1980s. Here values are formed and expressed in distinct opposition to the backroom dealings of secretive and undemocratic nations and institutions like the Soviet Union and OPEC. Although the saying in the pits was that "nobody is bigger than the market," media images suggested that this was a place of unparalleled individual freedom.[135]

The oil futures traders themselves, all of whom were Americans, also understood themselves to be fighting against OPEC's price setting power. As OPEC's pricing power continued to erode in 1985, one trader beamed, "Just the thought that you have this sort of control over as great a commodity as oil is tremendous. . . . It's just a super feeling. Especially since we can make the prices look bad and force the oil countries to lower their prices."[136] By collectively speculating a lower price for oil, traders were able to manifest those expectations into prices that made OPEC's "look bad." One exchange official expressed similar self-awareness, telling the *Guardian* that "the whole market went quiet. . . . Then, everybody clapped" when the price of WTI dropped more than 60 percent to below $10/barrel in April 1986. These NYMEX traders and officials took pride in their geopolitical role. As one commodity analyst noted about the generally patriotic sentiment of the exchange floor, "OPEC dared to restrict our mobility, our independence. Americans find that hard to forgive."[137] It seemed, for a time at least, that the turn to the market meant a return to cheap oil. America's independent producers—many of whom briefly reduced exploration budgets in the mid-1980s—may not have celebrated America's return to low prices, but consumers and traders rejoiced.[138]

If futures markets thrived on volatility, then what guarantee did American consumers have that oil prices would remain low? Liberal environmentalists and conservatives alike had supported market pricing in the late 1970s, thinking that it would translate into higher prices to discipline unruly consumption.[139] As long as oil gluts existed and prices continued to slide, as they had since 1981, it was possible to believe that NYMEX oil traders liked volatility at lower prices rather than volatility at higher prices. They would therefore keep prices much lower than OPEC because price swings, according to the *Guardian*, were smaller and less risky at \$15/barrel than at \$30/barrel, which made NYMEX the "natural enemy of OPEC." OPEC preferred higher prices because it meant higher profits, but NYMEX preferred lower prices because it meant less-risky profits.[140]

This point is crucial. Oil futures enjoyed such praise and fascination in the 1980s because they were perceived to be a driving force behind lower oil prices in the 1980s. For a US economy figured as an "oil consumer"—as the world's second-largest oil-producing nation had been since October 1973—cheap oil was an advantage that futures seemed to help to ensure. Although it is beyond the scope of this chapter to determine the precise role of futures in the oil price slide of the 1980s, there were several supply and demand factors also at play in oil prices. These included new access to Alaskan oil, riskier forms of oil extraction in the Gulf and other parts of the Outer Continental Shelf, North Sea oil discoveries, the Volcker Shock to interest rates, reduced consumption amid recession and new technology, and cooperation from European oil-producing countries that cut their posted prices and eventually abandoned the posted pricing system altogether. These factors, which are a blend of political, economic, and cultural forces, created the glut that weakened OPEC and made crude oil futures possible. As new oil flooded global markets at lower prices posted by oil-consuming nations, US commentators praised futures markets, bragged about their energy-conserving successes, and demanded the further reduction in prices to reflect the excess of supply, long-term ecological or climate considerations be damned. Despite these other geological and geopolitical factors rooted in Western dominance of global economic activity, the falling price of oil was figured as a free market phenomenon. As Chevron chairman George Keller claimed, the collapse of OPEC would "let the market . . . decide, and the price will undoubtedly fall flat and hard."[141] As oil futures, imagined as a technology for lowering prices, appeared to fulfill their role, they were cheered as an example of the market at work, forcing secretive cartels to play by the rules of supply, demand, and price, even as the West sought to form a cartel of consuming nations and to use its significant productive capacities to shape the global oil market.

A degree of duplicity inflected the financial world's geopolitical rhetoric over oil futures and OPEC. Aside from enabling the US economy to grow by exchanging title to future oil, oil futures enriched US stock exchanges and some of their established traders immensely by enabling the financial system "to cash in on de-regulation."[142] Speculators in oil and other commodities pursued their own financial interests, generating fascination at prodigious sums won and lost. The exchanges became wealthy economic centers at the heart of a newly financialized US economy.[143] The introduction of its options contract on crude helped the once-struggling NYMEX to become the third-largest commodity exchange in the nation. The price of a seat on the exchange jumped from $20,000 in 1980 to $168,000 in 1986.[144] Along the way, more American investments and savings joined in the oil futures bonanza.[145] Such success, according to acting CFTC chairman Gary Seevers, proved that there was "a real need for trading" and that people were "making money" on futures.[146]

When global oil prices finally collapsed in 1986, the exchanges and the media celebrated oil futures as a success with tremendous growth potential.[147] Already by 1984, Arnold Safer declared the oil market finally to be "a competitive market" that had "become what the futures price is."[148] *Newsweek* boasted that "never has it been quite so clear: the price of oil is no longer determined by the wave of the OPEC scepter but by the manipulations of Adam Smith's 'invisible hand.'"[149] Likewise, the CBOT in 1986 published what one reporter called "a study with [a] forgone conclusion" that "the competitive, open outcry futures market was among the most liquid and efficient market-making arrangements ever devised."[150] Now, after more than a decade of uncertainty about the energy future and the viability of US consumer capitalism, oil futures markets were touted as a crucial part of a solution to bring prices down and restore US power and vitality. *Newsweek* called the decline of OPEC "a satisfying spectacle."[151]

Conclusion

Oil futures markets were well-established by the end of the 1980s, generating billions of dollars of profit through the circulation of paper barrels and, so it seemed, keeping oil prices agreeably low. The establishment of oil futures as a key solution to the energy crisis relied on neoliberal narratives about the nature of the energy crisis, risk, and futures speculation. These narratives, which circulated among financial institutions, traders, politicians, and the media, reinterpreted the energy crisis as a financial crisis rather than as a crisis of overconsumption and scarcity. They claimed that economic volatility was now a permanent feature of the American experience and insisted that only the creation of oil futures markets could help to weather the storms of a post–Bretton Woods global economy. Not only was the institutionalization of oil price speculation framed as a rational response to volatility, it also promised to end the energy crisis and restore US geopolitical power by taking pricing power back from OPEC. Given that only

12 percent of Americans in 1986 felt that they understood commodity futures well enough to explain them to someone else, it is clear that these narratives had limited popular appeal.[152] Tracking them, however, reveals the development of the grand narrative of neoliberal financialization among the elites who stood to benefit the most.

These narratives had particular power because they tied financialization directly to the energy crisis, which many leaders, citizens, and experts had interpreted as an existential threat to US consumer capitalism.[153] The energy crisis threatened to undermine mass consumption—not to mention US hegemony—through inflation, high oil prices, and energy scarcity. By promising restoration rather than limits in the 1980s, a narrative congruent with the rhetoric of the Reagan administration, oil futures advocates painted an appealing picture of rational markets, lower prices, and energy abundance. The fossil economy seemed safe and secure under financialization. As the US federal government turned rightward, these narratives held sway where it mattered.

The legitimation of oil futures represented one among many developments in the longer process of financialization in the twentieth century, which included the normalization of stock investing and of consumer debt.[154] Each of these developments suggests the importance of viewing the history of neoliberalism, and other large-scale economic shifts, through the lens of cultural struggles over meaning.[155] Indeed, capitalism itself, as Jens Beckert argues, relies on "imaginaries of economic futures" for its functioning and continued existence.[156] For a country struggling to cope with the implications of the energy crisis for the future of its economy and way of life, oil futures contracts were perhaps the ultimate "instrument of imagination" that reconfigured price as anticipation and gave American markets the illusion of control over oil. By financializing oil futures, though, the exchanges also planted the seeds of future crises.

Enduring Crisis

The financialization of oil futures empowered "affective facts" and brought new sources of instability into the dynamics of global oil flows.[1] By the end of the 1980s, futures that might never happen could manifest in present prices, shaping the economy and extending crises. When Iran bombed a Saudi oil tanker in 1984, oil futures prices spiked and raised concerns that the United States and Saudi Arabia could be pulled into the Iran-Iraq War.[2] The *Wall Street Journal* described subsequent trading on NYMEX as "sometimes frantic" as its crude oil contract set a record volume on the possibility of a broader conflict in the Middle East.[3] Oil prices again surged in 1986 in several instances of heightened tensions in the Middle East and when the Soviet Union shut down sixteen nuclear power plants for safety inspection.[4] Even news about OPEC meetings could send prices upward as "pandemonium and panic" hit the trading floor.[5] Yet because these episodes seemed like blips in a downward oil price trend, they did not stir much public outcry. That would change during the 1990 Persian Gulf War crisis.

On August 2, 1990, Iraq invaded its small, oil-rich neighbor, Kuwait. The invasion sparked the first geopolitical crisis of the post–Cold War era and the first war to be waged entirely over the control of global oil production. Iraq was heavily indebted to Saudi Arabia and Kuwait after its war with Iran, and falling oil prices were diminishing the country's ability to meet its debt obligations, leading Iraq's dictator, Saddam Hussein, to pressure OPEC to increase oil prices to help replenish Iraqi state coffers. In order to justify his ambition to turn Iraq into a regional nuclear power, Hussein also declared that Kuwait had been unjustly severed from Iraq by Western imperialists and accused Kuwait of stealing Iraqi oil and conspiring with the United Arab Emirates and the United States to depress oil prices.[6] Counting on US neutrality in the interest of maintaining Iraq as a secular regional power to balance Iran, Hussein amassed troops along the border with Kuwait in late July 1990. Even then, few observers thought an invasion to be likely.

Iraq's invasion and annexation of Kuwait gave it control of 20 percent of the world's proven oil reserves and the potential for more should it invade Saudi Arabia. As political scientist Walter Goldstein remarked, Hussein came close to achieving "unlimited authority over oil prices, Western security and the nuclear war potential emerging in the Middle East."[7] The United States was particularly concerned because by 1990 imported oil accounted for 50 percent of its oil consumption. Acting in response to the security implications of Iraqi oil power and eager to assert its authority to lead a "New World Order" after the Cold War, the United States and a broad coalition including France, Canada, Saudi Arabia, Egypt, and Syria intervened in defense of Kuwait.

Oil futures skyrocketed in response to rumors and news of Iraq's invasion of Kuwait. The price of oil jumped 75 percent between mid-June and early August.[8] Throughout August and September 1990, oil prices fluctuated in response to public statements made by Hussein and President George H. W. Bush and news of military maneuvers, as well as wild rumors about phantom attacks and the dreams of dictators. In September, news that Hussein had "threatened to destroy" Saudi oil infrastructure pushed oil futures prices to a record $40/barrel, double their pre-crisis levels. Prices eventually steadied in December as Saudi Arabia and other nations increased production, but uncertainty remained about what would happen should the United States initiate a war. When the United States finally launched Operation Desert Storm on January 17, 1991, oil futures spiked by $10/barrel before falling to settle at around $20/barrel.[9]

Although short-lived, the panic over oil prices in those five months revealed the tensions inherent in extending neoliberal futures markets to the lifeblood commodity of the US and global economies. As one newspaper noted, there was "panic, greed and rising anarchy at the gas pumps and in the pipelines" as prices went up across the United States.[10] *The Times* observed that oil traders on London's International Petroleum Exchange could "taste blood" in the pits as they scrambled to hedge and profit from this jolt of uncertainty in the Mideast.[11] Rather suddenly, the futures market contracts that had promised risk management at lower prices were betraying the ends of US security and hegemony that they had been established to serve.

The skyrocketing prices and persistent volatility that marked the Persian Gulf crisis led many observers to wonder if this wasn't the "third oil shock" and to speculate about its potential consequences for an American economy already teetering on the brink of recession in the early 1990s.[12] The shocks of the seventies generated panic buying, price spikes, and widespread speculation about the nation's economic and energy prospects, but they bore some relation to physical shortages of oil, even if those shortages were more political than geological. The shock of the Persian Gulf crisis differed in that it was an entirely speculative phenomenon unrelated to material shortages. Oil prices on the futures markets doubled in August and September 1990 in the expectation of war with Iraq despite relatively strong "fundamentals" of supply and demand based on large private and government oil reserves in the United States and relatively weak demand from struggling industrial economies. Although a war did break out in 1991, the expectation of widespread oil shortages never materialized, except in the futures prices of late 1990.

The crisis demonstrated that speculative price fluctuations driven by a constant stream of information and rumor could shake the economic stability and security of the United States. Now, the material economic conditions of the present, in this case the realities of oil supply and demand, seemed to matter less than the expectations of traders about potential economic conditions in the future. As one expert told the US Senate during oil price hearings in October and November 1990, prices in futures and cash markets had become, "in effect, a consensus judgment reflecting the aggregate market expectations of both current and prospective market conditions."[13] As volatile market expectations persisted, media commentators, politicians, and economists turned against the oil futures markets and traders that they had initially praised for lowering the price of oil by undercutting OPEC. They agreed that Hussein's actions and months of US government deliberations had injected uncertainty in futures markets but chastised traders for irrationally spiking prices without the justification of an actual shortage.[14] The theory and rhetoric of oil futures markets throughout the 1980s had framed them as an ordered and rational response to uncertainty designed to reduce volatility and lower prices. The Persian Gulf crisis overturned those arguments as criticisms abounded that traders were acting emotionally or irrationally, turning fear and rumor into extravagant price swings and manufacturing price volatility where there should have been none.[15] As one trader worried aloud: "Normally, we love volatility, but this is insane."[16]

Most charges of speculative irrationality pointed to the market's apparent oversensitivity to news and rumor. It seemed that any scrap of information about the possible intentions or actions of national leaders could produce gains or losses of millions of dollars in oil futures value. For instance, prices fell in mid-August 1990 on speculation that a visit between President Bush and the king of Jordan might lead to a peace settlement, and then fell again when Saudi Arabia announced that OPEC might raise its production to 22.5 million barrels per day.[17] Prices tumbled in October when Hussein publicly called for dialogue and when Bush addressed the United Nations in a speech that traders interpreted as conciliatory.[18] But prices also skyrocketed on rumors of US troop movements in the Middle East in early August and then again in November when Bush publicly refused to "appease" Hussein.[19] And, perhaps most infamously, rumors that the Prophet Mohammed appeared to Hussein in a dream telling him to withdraw from Kuwait helped to spur an oil price drop in mid-October 1990.[20] Futures markets seemed to move without "economic justification."[21] Some even argued that the irrational panic of futures traders benefitted OPEC by ceding control over their market to Saddam Hussein. As one trader claimed, the market was now "tied to Saddam Hussein. If he sneezes, it will move."[22] Indeed, the potential market power of the dictator's sneeze was an oft-repeated trope in media coverage, underscoring the rapidity and irrationality of oil price movements during the crisis.[23] Value seemed to be won and lost thoughtlessly, on a whisper or a sneeze rather than reason.

Explanations for what was happening to oil futures prices contradicted neoliberal notions of futures markets as efficient information processors that reduced volatility. Uncertainty, which oil futures had been established to manage, was now the problem. As one British analyst argued, the oil markets did not have "all of the facts" about politics and military action that were potentially relevant to oil prices, so they had turned into "risk management without knowing all the risks."[24] Indeed, a broad consensus emerged that traders were "trading headlines" because they did not know what was really going on in the Mideast.[25] NYMEX had claimed in the 1980s that its crude oil contract was a response to volatility, but now it seemed clear that NYMEX had institutionalized and legitimized uncertainty about the future as a factor in oil prices. When there was little public information to process, markets could become volatile very quickly, causing economic panic and angst.

These discourses of market irrationality revealed persistent anxieties about the relation between the real economy and speculation, which have surfaced reliably throughout the history of futures trading in the United States. If present supplies and reserves were sufficient, why should present prices reflect a small chance of future scarcity? Aside from fear and a lack of information, some observers pointed to the fact that traders were not oil experts and had no abiding interest in it. They traded "paper barrels" for money and embodied a growing gulf between the real economy and the pursuit of money that was becoming a serious problem. One reporter concluded from his experience with futures trading that it was "a world where money is not the biggest consideration—it's the only consideration. With millions to be made or lost in seconds, there is no time for reflection on politics, morality, or the human condition. It doesn't even matter what the commodity is."[26] Reports about the pits during the crisis suggested that the pursuit of profit divorced from knowledge of the actual production, consumption, or trade of the commodity had severed its prices from real world conditions.[27] Speculation, as in the late nineteenth century, seemed to exist only for its own sake and the profits that it could generate for a privileged group of traders. This, of course, was the logical outcome of financialization; the Persian Gulf crisis merely revealed the economic violence inherent in it.

Congressional hearings in October 1990 revealed the crux of the crisis for policy makers: by raising the price of oil, futures markets were no longer supporting US economic hegemony, and so they were dismissed as vessels for irrational speculation detached from a reality presumed to support the desires of US oil consumers. Failing to apprehend that financialization entailed the equation of affect with value, Senator William Cohen (D-Maine) complained that "the fear of war has generated billions of dollars in profits, not the forces of supply and demand."[28] Senator Joe Lieberman (D-Connecticut) reiterated this position, saying that there was "absolutely no supply and demand reason for the price of oil to shoot up so high so quickly."[29] Instead, he blamed "speculators and profiteers" for exploiting the "psychological climate" of the Persian Gulf crisis with no concern for "economic reality."[30] There it was: speculation divorced from economic truth distorted prices to the detriment of US interests. Lieberman wondered if futures traders knew enough about the security of oil fields, military activities, and other government information to make informed and rational decisions. Was it even possible to have a rational market for a commodity of such geopolitical importance when it could not access all of the relevant information? Yet Lieberman missed the fact that "economic reality" in futures markets amounted to consensus expectations rather than any notion of "real" supply and demand relations. As one financial consultant let slip during the hearing: "Truth is irrelevant. How the market is going to perceive it is all that matters."[31] Markets had finally reached a point of abstraction such that the limits to growth did not apply when the economy could manufacture value from rumor, expectation, and fear. Whether this dynamic served "US interests" depended on the direction of the price. Nevertheless, it revealed the ethical and epistemological vacuity of markets unmoored from other goals and limiting institutions.

As the Persian Gulf crisis wore on, some economists and exchange officials defended futures trading by claiming that it had actually prevented oil prices from fluctuating as wildly as they otherwise would have. As prices calmed, they called the crisis a temporary aberration in a downward trend. NYMEX president Patrick Thompson claimed that the futures markets did not amount to "unfounded frenzy" but rather "legitimate commercial planning" in the allocation of resources.[32] More pointedly, NYMEX chairman Lou Guttman declared at an American Petroleum Institute brunch: "I didn't hear anybody complaining when the price went from $30 a barrel to $10 in the mid-80s that the Nymex was displaying prices. We don't make the prices, we just make them known."[33] Guttman was, in a sense, correct. Few people paid attention to the exchanges as oil prices dropped, but they were quick to condemn when the trend reversed. But his comment also revealed the fundamental ambiguity of futures markets that caused so much anxiety in the first place. If the markets just "revealed" prices but was not responsible for them, where did prices come from? Some commentators had been insisting that prices represented aggregate expectations about the future and that those expectations were cultivated by individual traders and institutions, who for the last few months had been responding to drips of information with wild price swings. According to NYMEX, these swings were merely revealed prices. For many others, they represented the dangerous "animal spirits" of the markets that could strike at any time.[34]

The speculative volatility of oil futures prices in late 1990 suggests that neoliberalism had trouble integrating oil into its market vision of the world. The materiality of oil resisted smooth integration into the financial system. Its geographic distribution meant that the United States depended on political stability outside of its borders to ensure economic stability within. Its geologic nature means that oil cannot be grown, like wheat or corn, and reserve estimates, as neoliberals often recited, are inherently uncertain and subject to change. The supply of oil itself, regardless of the nation that owns it, is inherently uncertain and limited. It cannot be counted with certainty and it cannot be multiplied. It can only be found, managed, and negotiated. Finally, as a relatively cheap and convenient source of energy, still the lifeblood of the global economy in 1990, oil was too important to the economic vitality of the United States in terms of the functioning of its economy and everyday life for millions of people to have its value fluctuate wildly. Under financialization, the generation of a crisis only required the possibility of a crisis in one part of the world. Thus, the rhetoric of NYMEX was right in a certain sense: under a regime of neoliberal financialization, the global economy was more volatile and uncertain, and crisis management was more important than ever before. But it failed to acknowledge its role in the institutionalization of speculation that made crisis easier to achieve and to magnify through expectation.

A Durable Crisis

The energy crisis persists. It is durable, able to incorporate new events and concerns into an ongoing sense that something is wrong with the way that modern societies extract and expend power from nature. Energy crisis narratives cropped up again after the petrodollar-funded terrorist attacks of 9/11 and the return of US military intervention in Iraq in 2003 renewed concern about the carbon foundations of US capitalism. A fresh wave of peak oil discourse swelled in North America and Europe in the aughts, reviving the scarcity paradigm of the seventies to argue for a new transition to renewable energy.[35] Much of this work used a familiar language of apocalyptic futurity in which the seventies energy crisis stood as a dire warning that had been ignored. Other responses renewed the seventies critique of US oil dependence, sometimes in terms of oil scarcity, other times due to the cost of oil, and most often because US oil dependence required violent interventions around the world to secure its supply.[36] Even President George W. Bush, the progeny of an oil family, renewed the language of national oil "addiction" in his 2006 State of the Union Address in which he vowed to begin the process of reducing oil imports from the Middle East by 2025 with a mix of ethanol and other energy forms.[37] Thirty years after the oil embargo exposed the fragility of US petroculture and sparked a decade of anxious introspection that gave way to the financialization of US hegemony, the idea of an energy crisis continued to animate US political culture. But it also remained a flexible crisis. Several critics of peak oil have reframed the ongoing energy crisis as a crisis of abundance at low prices that threatens to worsen environmental and social degradation.[38] Given the spread of fracking, tar sands, and shale oil that rejuvenated US oil production during the Obama and Trump presidencies, it has become clear that scarcity is not the only way to have an energy crisis, particularly under the shadow of climate change.[39]

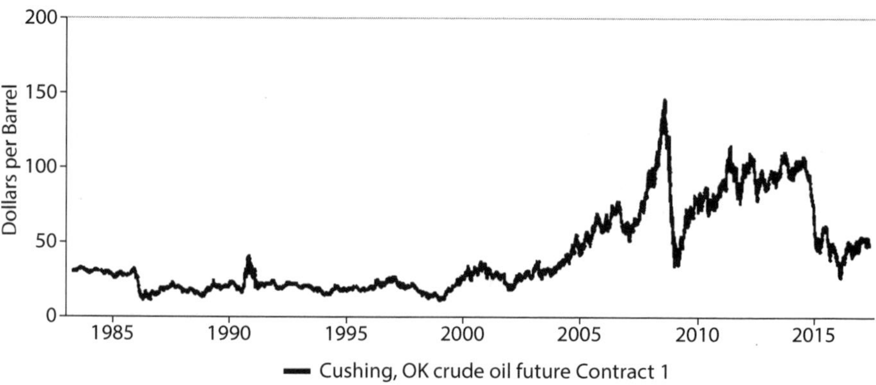

Figure 13. The 1990 oil futures spike paled in comparison to 2008. Source: Data from US Energy Information Administration.

The return of peak oil and dependence discourse was also fed by the rising price of oil on the spot and futures markets after the US invasion of Iraq in 2003. Between 2003 and 2008, NYMEX oil futures prices rose from around $30/barrel to a peak of $145/barrel, shooting up precipitously in 2007 and 2008 from $51/barrel to its 2008 peak.[40] These record prices, which coincided with the 2007–08 financial crisis and ensuing Great Recession, also fueled criticism of the speculative economy. It seemed to be the 1990s again for several months as a chorus of critics, with a Bush in the White House conducting a war in Iraq, expressed outrage that oil prices bore no relation to the real dynamics of supply and demand thanks to index traders and other kinds of speculators. The Department of Homeland Security articulated the importance of the oil economy to national security by holding hearings about the role of speculation in the unprecedented price of oil.[41] Even as some defenders, like Philip Verleger, blamed the government for distorting the market by filling the Strategic Petroleum Reserve, anxiety about the effects of speculation ran high at the hearings.[42]

Unlike the 1990s, however, claims of crisis in 2008 pointed simultaneously to speculation and scarcity discourses of peak oil, even though the crisis itself was largely financial, rooted in unmet expectations about the relation between present and future value.[43] These manifestations of energy crisis, which succeeded one another from the 1970s to the 1980s when financialization-supplanted scarcity came to coexist as crises of energy multiplied under intensified financialization and energy consumption.[44] As oil prices fell in the wake of the Great Recession and the "Shale Revolution" (fig. 13), oil futures speculation exceeded its financial form to resonate once again with two kinds of 1970s scenarios: large-scale environmental degradation and potential energy frontiers driven by technology, though presently we continue to rely more on alternative carbon than the alternatives to carbon that their supporters promise will forestall limits forever.[45] Boosters are always optimistic about what lies to the right of the graph.

Conclusion

At the beginning of *Anti-Crisis*, Janet Roitman asks, "How did crisis, once a signifier for a critical, decisive moment, come to be construed as a protracted historical and experiential condition?"[46] Part of her answer is to explore the constitution of crisis as itself an object of knowledge that structures social scientific inquiry and historical narratives in the absence of transcendental belief. In this book, I have taken up Roitman's challenge to think more critically about the power of crisis to shape our stories about the world by exploring how the construction, expansion, and evolution of the 1970s energy crisis ramified across time and worldviews to become a permanent feature of our sociopolitical landscape. Scholars often rely on crises as generative moments for inquiry because they lay bare the power relations and assumptions that animate the past and present.[47] Certainly, the dynamics of the COVID-19 crisis that began in 2020 will inform research and calls for social change for years to come. Although I have worked within this tradition by exploring the resonance between the sociopolitical economy of the energy crisis and neoliberalism, I have sought to do so reflexively by attending to the ontology and epistemology of the crisis and the political work that it accomplished, not the least of which was to survive the end of the oil embargo that thrust it into public discourse in 1973. It was an assemblage that proved to be flexible and durable, producing myriad effects.

I have traced a cultural history of the 1970s energy crisis to show how it remade the United States. The crisis reframed the US economy as an oil-consuming economy and enabled its financial reconstitution through speculation. It helped to expose the ecological consequences of carbon dependence and generated a range of ideas about how to address the problems of modernity: population, pollution, and alienation. As financialization supplanted ecological critique, the energy crisis helped to further remake the marine and terrestrial ecology of the United States—and the world—through the discourses of dependence that it produced and the threat to US hegemony that they posed. By obfuscating the ecological angle on the energy crisis while amplifying the desirability of energy independence, the energy crisis underwrote the expansion of "tough oil" development that continues to spread over land and sea in the name of security and economic necessity. Culturally, the energy crisis became a contested site over the meaning of capitalism, growth, and freedom that has oriented many people and politics toward entrepreneurial individualism, risky speculation, and petro-populism. It also posed anxious questions about the viability of the "American way of life" that were eventually "resolved" through market reverence. The energy crisis therefore revealed and, eventually, intensified the role of the United States in the dynamics that constitute the Anthropocene.

In *Carbon Democracy*, Timothy Mitchell notes the constructed nature of the energy crisis; it involved the conflation of different epistemological and ontological realms under a single field of energy. I have argued in this book that inasmuch as it enveloped America's economy, culture, and environmental politics through the analytic of energy, the energy crisis was also a central factor in the neoliberal reconfiguration of US political culture. Building on the insights of Huber, Jacobs, and others about the role of the crisis in popularizing ideals about energy markets free from government intrusion, I have shown how the energy crisis generated competing *narratives* that rendered the market more appealing, including petro-populist resistance to government regulation of automobility and neoliberal arguments in favor of markets to solve the crisis. Always a speculative phenomenon, the energy crisis built on the anxieties of the Cold War to help embed speculation in political life, bolstering its power to shape the present. Energy crisis narratives posited an abstraction, energy, as an urgent problem for which only freer markets would suffice. In so doing, the energy crisis has persisted as an arena of cultural and political contestation—just one more crisis in the ongoing crisis of neoliberalism that offers only more markets to cope with insecurity.

In recent years, an even more expansive crisis has coalesced that had only begun to enter public conversation when the energy crisis first emerged: climate change. Its complexity and scope have induced Timothy Morton to suggest the concept of a "hyperobject" to describe it and other entities that exceed our existing categories.[48] But as a space of discourse and action, climate change is similar to the energy crisis. Rooted in the pursuit of growth and reliance on carbon energy at the root of the energy crisis, the climate crisis is speculative, always interpreting current disasters as portents of future possibilities that threaten biological and democratic life as it currently exists. The climate crisis also raises complex concerns about security, dependence, population, and international cooperation that are inseparable from the energy crisis, even as they exceed it. It is contested and flexible, supporting renewed and more urgent calls for an energy transition, but also for the establishment of market solutions and irrevocable interventions in the ecology of the earth's atmosphere. The climate crisis poses a limit to growth, akin to the energy crisis, and is met with proposals to exceed those limits. But in their seeming intractability, the climate and energy crises pose fundamental questions about human agency, values, structures, and virtues, about politics and ethics. How should we live under these conditions? What ways of thinking, relating, and being in the Anthropocene might be more suited, at multiple scales, to the conditions that support life and human flourishing? Which scales are the most important and humane? Can we think about energy beyond the terms of price, risk, and crisis? Can we imagine life after oil? To do so requires wrestling with the past and the categories that we have inherited from it that shape our present horizons.

Notes

Abbreviations

API	American Petroleum Institute
CONAES	Committee on Nuclear and Alternative Energy Systems
COWPS	Council on Wage and Price Stability
GFPL	Gerald R. Ford Presidential Library and Museum, Ann Arbor, MI
GSL	George A. Smathers Libraries
HTOP	Howard T. Odum Papers
JCPL	Jimmy Carter Presidential Library and Museum, Atlanta, GA
NARA	National Archives and Records Administration
NMAH	National Museum of American History
NYPLA	New York Public Library Archive
RAC	Rockefeller Archive Center
WSP	William E. Simon Papers

INTRODUCTION: **Energy in Crisis**

 Epigraph. Roitman, *Anti-Crisis*, 93.

 1. Mitchell, *Carbon Democracy*, 173.

 2. Yergin, *The Prize*, 602ff; Jacobs, *Panic at the Pump*, 49ff; Graf, "Making Use of the 'Oil Weapon,'" 185–208.

 3. Price-Smith, *Oil, Illiberalism, and War*, 18–20; Carlisle et al., *The Politics of Energy Crises*, 88.

 4. Jacobs discusses US federal government policies in detail.

 5. Carlisle et al., *The Politics of Energy Crises*, 128ff.

 6. Carlisle et al., 130.

 7. Yergin, *The Prize*; Venn, *The Oil Crisis*; Sargent, *A Superpower Transformed*; Clayton, *Market Madness*; McNally, *Crude Volatility*; Parra, *Oil Politics*.

 8. On the challenge of representing fossil fuels, see Szeman, *On Petrocultures*. On the relationship between culture, politics, and materiality, see Mitchell, *Carbon Democracy*, and Huber, *Lifeblood*.

 9. Nixon, "Remarks at the Seafarers International Union Biennial Convention."

 10. Petrocultures Research Group, *After Oil*, 9–10.

 11. See Nye, *Consuming Power*.

 12. It is worth asking if electricity generated from local solar panels is the same thing as electricity generated by burning coal.

13. White, *The Organic Machine*; Black, *Crude Reality*; Nye, *Consuming Power*; Jones, *Routes of Power*; Bakke, *The Grid*; Smil, *Energy and Civilization*.

14. On energy regimes, see McNeil, *Something New Under the Sun*. On energy systems, see Debeir, Delége, and Hémery, *In the Servitude of Power*.

15. Historians addressing this gap include Jones, *Routes of Power*, and Johnson, *Carbon Nation*.

16. Exceptions include Malm, *Fossil Capital*, and Jones, *Routes of Power*.

17. Johnson, *Carbon Nation*, xviii.

18. Jones, *Routes of Power*.

19. Huber, *Lifeblood*, 29.

20. McNeill and Engelke, *The Great Acceleration*.

21. Wapshott, *Keynes Hayek*, 226–246.

22. Gerstle, "The Reach and Limits of the Liberal Consensus," 55–56.

23. Borstelmann, *The 1970s*, 127–128.

24. Cowie, *The Great Exception*, 9.

25. Cowie, 154.

26. Cowie, 11, 19–24.

27. Cowie, 29.

28. Cowie, 16.

29. Williams, "Culture Is Ordinary," 9.

30. For useful summaries, see Janiewski, "Review," 60–67; Phillips-Fein, "Conservatism," 723–743. See also McGirr, *Suburban Warriors*; Lichtman, *White Protestant Nation*; Moreton, *To Serve God and Wal-Mart*; Phillips-Fein, *Invisible Hands*; Kalman, *Right Star Rising*; Flippen, *Jimmy Carter*; Sandbrook, *Mad as Hell*; Self, *All in the Family*; Fones-Wolf and Fones-Wolf, *Struggle for the Soul of the Postwar South*.

31. Carter, *From George Wallace to Newt Gingrich*; Self, *American Babylon*; Kruse, *White Flight*; Crespino, *In Search of Another Country*; Lassiter, *Silent Majority*; Mason, *Reading Appalachia from Left to Right*; Brown and Smith, *Race and Real Estate*. On the concept of racial formation, see Omi and Winant, *Racial Formation in the United States*.

32. Freund, *Colored Property*.

33. Lassiter, "Political History beyond the Red-Blue Divide," 760.

34. Jacobs, *Panic at the Pump*.

35. Bruce Shulman has recently labeled arguments focused on neoliberalism like the one I make in this book as a "neo-consensus" approach to post-1960s US history. Yet in its emphasis on the hegemonic cultural and political dynamics of fossil fuels and crisis, this book does not negate the importance of "partisan and ideological conflicts." See Bruce J. Shulman, "Post-1968 U.S. History: Neo-Consensus History for the Age of Polarization," *Reviews in American History* 47, no. 3 (September 2019), 479–99. (Quote is on page 480.)

36. Hohle, *Race and the Origins of American Neoliberalism*. In some ways, neoliberalism more fully economizes life than some forms of neoconservatism. In a book that dismisses the concept of mediating institutions, Melinda Cooper argues that neoliberalism and neoconservatism share significant ground in how they value family. Cooper's arguments are often insightful, but she overstates the significance of this connection and

dismisses the corrosive effects of the neoliberal paradigm on families, a point that many conservative critiques have developed. Cooper, *Family Values*. Cooper also misses how many of the "non-normative desires" that she posits as targets of neoliberal and neoconservative agreement on "family values" are fully compatible with global neoliberalism.

37. Dieter Plehwe, introduction to Mirowski and Plehwe, *The Road from Mont Pelerin*, 2.

38. McGuigan, *Neoliberal Culture*, 1–3; Dumenil and Levy, *The Crisis of Neoliberalism*, 1–10; Slobodian, *Globalists*; Stedman Jones, *Masters of the Universe*; Plehwe, introduction, 14. The last insight comes from Foucault, *The Birth of Biopolitics*, 120ff.

39. Foucault, "Lecture 10," in *Birth of Biopolitics*, 239–265; Philip Mirowski, "Postface: Defining Neoliberalism," in Mirowski and Plehwe, *The Road from Mont Pelerin*, 423–424. For more on neoliberal epistemology, see Walker and Cooper, "Genealogies of Resilience," 143–160.

40. Brown, *Undoing the Demos*, 9.

41. Harvey, *A Brief History of Neoliberalism*; Tadiar, "Life-Times in Fate Playing"; Cazdyn and Szeman, *After Globalization*; Panitch and Gindin, *The Making of Global Capitalism*; Heynen et al., "Introduction: False Promises," in *Neoliberal Environments*, 3.

42. See Slobodian, *Globalists*, and Huber, *Lifeblood*.

43. Cooper, *Life as Surplus*, 10.

44. Adams, Murphy, and Clarke, "Anticipation."

45. Cooper, *Life as Surplus*, 11.

46. On structures of feeling, see Williams, "The Analysis of Culture," in *The Long Revolution*, 48; Williams, *Marxism and Literature*, 132.

47. Ventura, *Neoliberal Culture*, 2–3.

48. Huber, *Lifeblood*, 23.

49. Stoekl, *Bataille's Peak*.

50. De Rycker and Mohd Don, introduction, 6–7.

51. Koselleck, "Crisis"; OED Online, "crisis, n.," September 2019, Oxford University Press.

52. Roitman, *Anti-Crisis*, 31. Roitman also indicts post-structuralism for relying on crises as the "starting point for narration," such as the crisis of the subject. Roitman, 34.

53. Roitman, 35.

54. Roitman, 17. See Koselleck, *Futures Past*, and Koselleck, *The Practice of Conceptual History*.

55. Roitman, *Anti-Crisis*, 11, 7.

56. My aim is to think through the historical ontology of the energy crisis and to extend the concept of assemblage, typically applied to material phenomena, to crisis. For a discussion of assemblages, see Murphy, *Sick Building Syndrome*, 11–16.

57. Stein, *Pivotal Decade*.

58. Zaretsky, *No Direction Home*.

59. Jacobs, *Panic at the Pump*; Fiege, *The Republic of Nature*, 358–402.

60. Stein, *Pivotal Decade*.

61. Mitchell, *Carbon Democracy*.

62. Huber, *Lifeblood*, xx.

CHAPTER 1: "Is America Running Out of Gas?"

The main title of this chapter is taken from "Slide: Economics: Is America Running Out of Gas?," Folder 13, Box 118, American Petroleum Institute (API) Photograph and Film Collection, 1860s–1990, Archives Center, National Museum of American History (NMAH).

1. *Time*, Jan. 21, 1974. See the cover story: "The Whirlwind Confronts the Skeptics," 20. See also Huber, *Lifeblood*, 116ff; "Some Doubt a 'Crisis' but the Price Rise Is Real," *Washington Post*, Dec. 26, 1973.

2. Potomac Associates, Gallup / Potomac Associates Poll: *State of the Nation, 1974*, Question 182, USGALLUP.74POTC.Q36, Gallup Organization (Cornell University, Ithaca, NY: Roper Center for Public Opinion Research, 1974).

3. Letter, William E. Simon to C. Roderick O'Neill, 14 Jan. 1974, Folder 20, Box 13, William E. Simon Papers (WSP), Gerald R. Ford Presidential Library and Museum (GFPL), Ann Arbor, MI.

4. Letter, William E. Simon to George W. Bramhall, 28 Jan. 1974, Folder 18, Box 1, WSP, GFPL.

5. Memo, James Schlesinger and Stu Eizenstat to the President, 10/9/78, Domestic Policy Staff Papers Stuart Eizenstat's Subject Files, Energy Bill [8], Box 199, Jimmy Carter Presidential Library and Museum (JCPL), Atlanta, GA. See also Mattson, *What the Heck Are You Up to Mr. President?*

6. Executive Office of the President—Energy Policy and Planning, *The National Energy Plan*, p. 9, Council of Economic Advisers George C. Eads' Subject Files, Energy Plan, Box 223, JCPL.

7. Letter, Daniel Yankelovich to President Carter, 06/07/77, Domestic Policy Staff Stuart Eizenstat's Subject Files, "Energy Bill [9]," Box 199, JCPL.

8. Roitman, *Anti-Crisis*, 3.

9. Michelle Murphy defines an assemblage as "an arrangement of discourses, objects, practices, and subject positions that work together within a particular discipline or knowledge tradition." It emphasizes the historical specificity of the perceptibility of new "qualities, capacities, and possibilities." Applied to the energy crisis, assemblage allows us to see a fuller range of discursive, temporal, and material phenomena that informed the apprehension and articulation of a crisis in the US energy system: the crisis cannot be reduced to scarcity, inefficiency, policy, or any other single cause. See Murphy, *Sick Building Syndrome*, 12, n17.

10. On time and energy, see Kern, *The Culture of Time and Space, 1880–1918*.

11. Koselleck, *Future's Past*, 236. On space of experience and horizon of expectations, see Koselleck, *The Practice of Conceptual History*, 110–113. My reading of Koselleck is indebted to Roitman, *Anti-Crisis*, 15–39.

12. Koselleck, *Future's Past*, 269.

13. Koselleck, *Futures Past*, 262–263, 274; Roitman, *Anti-Crisis*, 17.

14. Lakoff, "From Population to Vital Systems," 33–60; Masco, *The Theater of Operations*.

15. US Energy Information Administration, *Energy in the United States, 1635–2000*.

16. Black, *Crude Reality*, 5–147.

17. Jones, *Routes of Power*.

18. Wells, *Car Country*; Nye, *Consuming Power*.

19. US Energy Information Administration, "What Is U.S. Electricity Generation by Source," https://www.eia.gov/tools/faqs/faq.php?id=427&t=3.

20. Huber, *Lifeblood*, 29.

21. Huber, 30.

22. McNeill and Engelke, *The Great Acceleration*, 4.

23. Cohen, *A Consumers' Republic*, 193ff.

24. Jackson, *Crabgrass Frontier*, 232; Cohen, *A Consumers' Republic*, 195.

25. Cohen, *A Consumers' Republic*, 141.

26. US Bureau of the Census, *Statistical History of the United States*, 639; Hayden, *Building Suburbia*, 128–153.

27. Hayden calls suburbia a "sprawl machine." Hayden, *Building Suburbia*, xii.

28. On the political importance of suburbs: McGirr, *Suburban Warriors*; Kruse, *White Flight*; Avila, *Popular Culture in the Age of White Flight*; Lassiter, *Silent Majority*; Dochuk, *From Bible Belt to Sunbelt*.

29. May, *Homeward Bound*, 157–158. See also Rome, *Bulldozer in the Countryside*, 42–43.

30. May, *Homeward Bound*, 158.

31. Rome, *Bulldozer in the Countryside*, 65ff. For more, see Cooper, *Air-conditioning America*.

32. William H. Whyte Jr., "Budgetism: Opiate of the Middle Class," *Fortune*, May 1956, 133.

33. Jackson, *Crabgrass Frontier*, 11. See also Mackin, *Americans and Their Land*, 123–134.

34. Jackson, *Crabgrass Frontier*, 232–233. On the Federal Housing Administration, see Wells, *Car Country*, 256–257.

35. Quoted in Jackson, *Crabgrass Frontier*, 231.

36. Marling, *As Seen on TV*, 242ff.

37. Wells shows that these developments—increasing numbers of cars, energy extraction, road building, and urban decentralization—began earlier in the century but intensified postwar. He also maintains that suburban development drove the spread of cars, rather than the other way around, citing studies from the 1960s demonstrating that residential location drove choice of transportation. Wells, *Car Country*, 280.

38. May, *Homeward Bound*, 158; Jackson, *Crabgrass Frontier*, 246.

39. Wells, *Car Country*, 279.

40. Rome, *Bulldozer in the Countryside*, 42; Cohen, *A Consumers' Republic*, 123.

41. Seiler, *Republic of Drivers*, 71–72. The Interstate Highway System was as much of an ideological project as it was utilitarian. It supported US individualism against the collectivist threat of the Soviet Union and corporate conformity. Automobility was framed as "a constituent element of the American character" that could arrest its decline and demonstrate the vitality of American individualism to America's allies and enemies. For Wells, the construction of the interstate further encouraged suburbanization and car dependency. Wells, *Car Country*, 254–255.

42. Mandel, *Driven*.

43. Carter et al., *Historical Statistics of the United States*, 37–339.

44. US Energy Information Administration, "Electricity Explained: Electricity in the United States," https://www.eia.gov/energyexplained/electricity/electricity-in-the-us.php.

45. Nye, *Consuming Power*, 6.

46. Hamilton, *Trucking Country*.

47. Huber, *Lifeblood*, 83–85.

48. *The Graduate* (Dir. Mike Nichols, 1967).

49. Wells calls this "car country," meaning "places where car dependence is woven into the basic fabric of the landscape." Wells, *Car Country*, xxx–xxxi.

50. Carter et al., *Statistical History of the United States*, 718, 721.

51. Wells, *Car Country*, 148.

52. Carter et al., *Statistical History of the United States*, 718.

53. Baudrillard, *America*, 7.

54. Wells, *Car Country*, xxx.

55. Toffler, *Future Shock*, 35–44.

56. Toffler, 69–71. Toffler thought that so much automobile-enabled travel reshaped subjectivity too, noting that "travelers and nomads are not the same kind of people who stay put in one place."

57. Odum, *Environment, Power, and Society*, 115.

58. Experts on population and pollution in the 1960s added energy to their concerns in the 1970s and worried about how the energy and population crises would unfold on a global scale. Ehrlich and Ehrlich, *The End of Affluence*, 91–116.

59. US Energy Information Administration, "Coal Explained: Coal Imports and Exports," https://www.eia.gov/energyexplained/coal/imports-and-exports.php, and "Natural Gas Explained: Natural Gas Imports and Exports," https://www.eia.gov/energyexplained /natural-gas/imports-and-exports.php.

60. Mancke, *Squeaking By*, 15–16; Jacobs, *Panic at the Pump*, 17–19.

61. Jacobs, *Panic at the Pump*, 19.

62. Jacobs, 25–36.

63. McNally, *Crude Volatility*, 115ff.

64. McNally, 95; Maugeri, *The Age of Oil*, 49–50, 65–67, 86–87, 118.

65. Price-Smith, *Oil, Illiberalism, and War*, 14–16.

66. Yergin, *The Prize*, 519ff.

67. McNally, *Crude Volatility*, 120.

68. McNally, 127.

69. "Anxieties of Abundance" recalls Daniel Horowitz's book *Anxieties of Affluence*.

70. Mitchell, *Carbon Democracy*, 173, 177–181. Energy companies formed when oil companies vied for control of coal, natural gas, and nuclear power in the early 1970s, controlling, for example, 40 percent of US uranium reserves.

71. Contemporaries made similar observations: Bruce Netschert, "The Energy Company: A Monopoly Trend in the Energy Market," *Bulletin of the Atomic Scientists*, Oct. 1971, 13–17; "Energy Firms: Suspicion and Mistrust," *Chicago Tribune*, July 5, 1973.

72. Rabinbach, *The Human Motor*, 1–11.

73. The rise of "systems theory" is another key context, inseparable from the ecosystem concept and systems ecology. Lilienfeld, *The Rise of Systems Theory*.

74. Hutchison was influenced by cybernetics. Hagen, *An Entangled Bank*, 69. These ecologists differed from early twentieth-century energy economists like Frederick Soddy by making systems primary and energy an animating feature of them, rather positing energy as a more abstract animating life force linked to bodies and labor. See Bramwell, *Ecology in the Twentieth Century*, 64–91.

75. Golley, *A History of the Ecosystem Concept*, 8.

76. Lindeman, "The Trophic-Dynamic Aspect of Ecology," 400. Lindeman was linking succession and energetics. The former was the dominant concern of pre-WII ecology, and the latter characterized ecological research in the Atomic Age. Hagen, *An Entangled Bank*, 93. See also Cook, "Raymond Lindeman and the Trophic-Dynamic Concept," 22; Kingsland, *The Evolution of American Ecology*, 188; Golley, *A History of the Ecosystem Concept*, 153–156.

77. Cook, "Raymond Lindeman and the Trophic-Dynamic Concept," 25.

78. Mitman, *The State of Nature*, 140–141.

79. See Hagen, "Eugene and Howard Odum."

80. Kingsland, *The Evolution of American Ecology*, 189–199; Hagen, *An Entangled Bank*, 122.

81. Odum, "Energy Flow in Ecosystems"; Odum, "The Ecosystem, Energy, and Human Values."

82. Odum, "Energy Flow in Ecosystems," 16.

83. Hagen, *An Entangled Bank*, 125.

84. Odum, *Environment, Power, and Society*, 1–2.

85. "The Nature of Energy," *New York Times*, Dec. 5, 1973.

86. White, "Energy and the Evolution of Culture," 335; White, *The Science of Culture*, 335.

87. White, *The Science of Culture*, 362, 369. White drew on the work of the German chemist Wilhelm Ostwald. See Stewart, "Sociology, Culture and Energy."

88. White, "Energy and the Evolution of Culture," 350.

89. Boyer, "Energopower," 311.

90. Boyer, 311.

91. Cottrell, *Energy and Society*, vii.

92. Cottrell, 198–99, 239–240; Nikiforuk, "The Social Price of Energy."

93. UNESCO funded a report with similar premises and conclusions about the relation between humans and energy: Egerton, "Energy in the Service of Man."

94. Farish, *The Contours of America's Cold War*, xii. Farish notes how, during the Cold War, different spaces were "defined, collectively, in strategic terms," an insight that applies to energy resources as well.

95. Harry S. Truman, "Letter to William S. Paley on the Creation of the President's Materials Policy Commission," online by Gerhard Peters and John T. Woolley, The American Presidency Project, https://www.presidency.ucsb.edu/node/230573. The Paley Commission was officially titled the "President's Materials Policy Commission," directed by William S. Paley, the head of CBS. Its report was titled "Resources for Freedom."

96. President's Materials Policy Commission, *Resources for Freedom: Summary of Volume I*, 5–6.

97. President's Materials Policy Commission, *Resources for Freedom: Summary of Volume I*, 2. The report recommended maintaining reserve oil capacity in the event of war. Such capacity evaporated in 1972.

98. Oil companies supported the commission's recommendation to keep a tax structure favorable to domestic exploration. "Oil Men Cheered by Federal Study," *New York Times*, July 14, 1952.

99. Mitchell, *Carbon Democracy*, 177, n10.

100. President's Materials Policy Commission, *Resources for Freedom: Summary of Volume I*, 65ff. Reporting on a similar study by the National Security Resources Board, the *Wall Street Journal* suggested both studies did not offer solid grounds for optimism on the future of energy resources because neither advocated nuclear power. "The Slammed Door," *Wall Street Journal*, Dec. 26, 1952. Media accounts from the 1960s also framed the Paley report timidly. "U.S. Resources Held Adequate," *New York Times*, April 17, 1966.

101. President's Materials Policy Commission, *Resources for Freedom: Summary of Volume I*, 11.

102. Energy Resources Committee, "Message from the President to the Congress of the United States," in *Energy Resources and National Policy*.

103. The quantitative data from *Energy Resources and National Policy* informed the State Department's *Energy Resources of the World* (1949), another entry in the genealogy of energy resource surveys.

104. President's Materials Policy Commission, "Foreword and Acknowledgement," *Resources for the Freedom, Volume III*.

105. President's Materials Policy Commission, *Resources for Freedom, Volume III*, 1.

106. President's Materials Policy Commission, *Resources for Freedom: Summary of Volume I*, 37, 40. The report argued that oil company access to the OCS would be useful to national security.

107. One example is Scarlott, *Energy Sources*. For a discussion of postwar works on the potential of alternative energy systems, see Barber, *A House in the Sun*, 82–88.

108. Hubbert, *Energy Resources*, 2.

109. Hubbert, 2.

110. Hubbert, 15.

111. Schurr et al., *Energy in the American Economy*, 31. A second report followed three years later: Landsberg, Fischman, and Fisher, *Resources in America's Future*.

112. Resources for the Future, *U.S. Energy Policies*, 2.

113. Darmstadter, Teitelbaum, and Polach, *Energy in the World Economy*, 3.

114. Malthus, *An Essay on the Principle of Population*, 13–29.

115. Schlosser, "Malthus at Mid-Century," 468–471.

116. Robertson, *The Malthusian Moment*, 5–7.

117. Jevons, *The Coal Question*.

118. Schoijet, "Limits to Growth and the Rise of Catastrophism," 516–517. At this time many believed that nuclear energy would eliminate resource scarcity.

119. Boulding, foreword, x.

120. Vogt, *Road to Survival*; Osborn, *Our Plundered Planet*.

121. On their intellectual milieu of nuclear apocalypticism, see Jundt, *Greening the Red, White, and Blue*, 11–17.

122. Vogt, *Road to Survival*, 18.

123. Schlosser, "Malthus at Midcentury," 474.

124. Vogt, *Road to Survival*, 287.

125. Ordway, *Resources and the American Dream*.

126. Rome, "Give Earth a Chance," 528–534.

127. Connelly, *Fatal Misconception*, 115–154; Hamblin, *Arming Mother Earth*, 151–178; Robertson, *The Malthusian Moment*, 104–125. American films articulated fears about overpopulation and resource scarcity: *Silent Running* (Dir. Douglas Trumbull, 1972), *ZPG* (Dir. Michael Campus, 1972), *Soylent Green* (Dir. Richard Fleischer, 1973), and *Logan's Run* (Dir. Michael Anderson, 1976).

128. Sabin, *The Bet*, 1–4, 11–12.

129. Buell, *From Apocalypse to Way of Life*, 180–181.

130. Ehrlich, *The Population Bomb*, 131.

131. Hardin, "The Tragedy of the Commons," 1243.

132. David Inglis, "Nuclear Energy and the Malthusian Dilemma," *Bulletin of the Atomic Scientists*, Feb. 1971, 14–18.

133. Meadows et al., *The Limits to Growth*; Schoijet, "Limits to Growth and the Rise of Catastrophism," 518.

134. Hamblin, *Arming Mother Earth*, 176–177.

135. Richard Nixon, "Special Message to the Congress on Energy Policy," American Presidency Project, https://www.presidency.ucsb.edu/node/255364.

136. Carson, *Silent Spring*, 1–2.

137. Ehrlich, *Population Bomb*, xi, 3.

138. Hamblin, *Arming Mother Nature*.

139. Masco, *The Theater of Operations*, 6–7.

140. On the cultural dimensions of the bomb, see Henriksen, *Dr. Strangelove's America*.

141. Masco, *The Theater of Operations*, 47.

142. Kahn, *On Thermonuclear War*.

143. See online "Energy Crisis" exhibition, National Museum of American History (NMAH), https://americanhistory.si.edu/american-enterprise-exhibition/consumer-era /energy-crisis.

144. I use "construct" in the sense of building something. The response of the Moline Illinois Dispatch to the assemblage of energy, population, and pollution crises suggests the difficulty of apprehending the energy crisis and the importance of belief. The paper said: "It is difficult for those of us living in the bountiful and blessed Midwest to comprehend what is happening, but the disturbing facts cannot be denied nor can the difficult decisions that are implied." The statement suggests a gulf between everyday existence and the crisis discourses of the period. "The Sunday Dispatch: A Global Report," 08/03/1980, "Global 2000[2]," Box 34, Domestic Policy Staff Natural Resources Ward/Schirmer, JCPL.

145. Landsberg, Fischman, and Fisher, *Resources in America's Future*, 30–32, 4.

146. Gaucher, "Energy Sources of the Future for the United States," 125. Gaucher foresaw the eventual exhaustion of present forms of energy, but only by the year 2200. Another example of modernist faith in technological progress: Brown, Bonner, and Weir, *The Next Hundred Years*, 4.

147. Hubbert said: "One of the most disturbing ecological influences of recent millennia is the human species' proclivity for the capture of energy, resulting in a progressive increase in the human population." The resulting population growth rate, he said, was "not normal." Hubbert, "Energy from Fossil Fuels," 104–105.

148. Hubbert, *Nuclear Energy and the Fossil Fuels*, 134–138.

149. Hubbert, *Energy Resources*, 127, 134–140.

150. Deffeyes, *When Oil Peaked*.

151. The energy crisis sapped Hubbert's optimism about nuclear power. Hubbert, "The World's Evolving Energy System," 1026–1027.

152. Hubbert, *Nuclear Energy and the Fossil Fuels*, 36–37.

153. Hubbert, *Energy Resources*, and "The Energy Resources of the Earth," *Scientific American* 225, no. 3 (Sept. 1971): 60–70.

154. *Life* Special Double Issue: "Into the 70s," Jan. 9, 1970.

155. Robert L. Heilbroner, "Priorities for the Seventies," *Saturday Review*, Jan. 3, 1970, 17–19, 84.

156. "The Storms of the 70s," *The Nation*, Jan. 12, 1970, 6–8.

157. "Some Burning Questions about Combustion," *Fortune*, Feb. 1970, 130–131.

158. "Energy: Oil's Future," *Forbes*, Jan. 1, 1970, 154–156.

159. Production did not surpass that peak until well into the Shale Revolution, in 2018: US EIA, "U.S. Field Production of Crude Oil," https://www.eia.gov/dnav/pet/hist /LeafHandler.ashx?n=PET&s=MCRFPUS2&f=A.

160. Huber, *Lifeblood*, 101–102.

161. "Northeast Oil Crisis Feared," *New York Times*, Feb. 7, 1970.

162. "Power Crisis: A Shortage of Energy Threatens Industries and Homes Across U.S.," *Wall Street Journal*, June 2, 1970.

163. "Scrambling for Oil: Supply of Industrial Fuels Grows Tighter, Adding to Fear of Energy Woes This Winter," *Wall Street Journal*, July 31, 1970.

164. "Scrambling for Oil," *Wall Street Journal*, July 31, 1970.

165. Matusow, *Nixon's Economy*, 241–275; Lifset, "A New Understanding of the American Energy Crisis of the 1970s."

166. Matusow, *Nixon's Economy*, 245, 137; Kohn, "The Oil Import Question."

167. Richard Nixon, "Special Message to the Congress on Energy Resources," American Presidency Project, https://www.presidency.ucsb.edu/node/240214. The oil industry lobbied intensely for more access to federal land: Series 4: Slides 1970s-1980s, Box 74, 79, API Photograph and Film Collection, 1860s-1990, Archives Center, NMAH.

168. Jacobs, *Panic at the Pump*, 30–36, 47–48.

169. Nixon, "Special Message to the Congress on Energy Resources."

170. On business and conservative resentment of environmental regulation, see Turner and Isenberg, *The Republican Reversal*.

171. The OCS and Alaska were early targets within the federal government as well. Philip L. Essley Jr., "Methods of Increasing Crude Oil Supply," folder Oil—Nov. & Dec. 1973, Box 1, Records of the Department of the Treasury, Record Group 56, NND 32191, National Archives and Records Administration (NARA), College Park, MD.

172. National Petroleum Council, *U.S. Energy Outlook: An Initial Appraisal*, 4–5. The report did not consider solar or any other renewable form other than hydro in any significant detail. Similar reports include: US Federal Power Commission, *The 1970 National Power Survey*; Dupree and West, *United States Energy Through the Year 2000*; Statistical Office of the United Nations, *World Energy Supplies*; Darmstadter, *Energy in the World Economy*; National Petroleum Council, *U.S. Energy Outlook* (1972).

173. API president Frank Ikard quoted in Stephan P. Potter, "The American Petroleum Institute: An Informal History (1919–1987), 1990, p. 46, Folder 1, Box 116, API Photograph and Film Collection, 1860s-1990, Archives Center, NMAH.

174. Potter, 51.

175. "Acute Energy Gap Seen by 1985," *Christian Science Monitor*, Aug. 8, 1972; Derek P. Gregory, "The Hydrogen Economy," *Scientific American* 228, no. 1 (Jan. 1973): 13; "Energy Gap," *New York Times*, Nov. 11, 1973; "Silver Lining for the Energy Gap," *Wall Street Journal*, Dec. 12, 1973; "Closing the Energy Gap," *Washington Post*, Feb. 2, 1977.

176. "Oil Group Warns on Ecology Issue," *New York Times*, Nov. 14, 1972. See also "Oil Executive Urges Energy Conservation," *Chicago Tribune*, Nov. 25, 1972.

177. "The Crisis behind the Crisis," *Chicago Tribune*, June 21, 1973; "Selling the Energy Crisis," *New York Times*, Apr. 1, 1973.

178. "Start Rationing Now, Oilmen Advise," *New York Times*, Nov. 14, 1973.

179. "The Crisis behind the Crisis," 20.

180. Some did see the energy crisis as a technical management issue: Rodgers, *Brown-Out*.

181. John Noble Wilford, "Nation's Energy Crisis: It Won't Go Away Soon," *New York Times*, July 6, 1971.

182. John Noble Wilford, "Nation's Energy Crisis: Is Unbridled Growth Indispensable?," *New York Times*, July 8, 1971.

183. Peter Borrelli, introduction to Holdren and Herrera, *Energy*, 9; Paul R. Ehrlich and John P. Holdren, "The Energy Crisis," *Saturday Review*, August 1, 1971, 50–51.

184. Holdren and Herrera, *Energy*, 15, 21–24.

185. Rocks and Runyon, *The Energy Crisis*.

186. Rocks and Runyon, 176–177. Other early treatments include Garvey, *Energy, Ecology, Economy*; Hines, *Environmental Issues*; Harrison, *Earthkeeping*.

187. G. Evelyn Hutchison, "The Biosphere," *Scientific American*, Sept. 1970, 45–53.

188. S. Fred Singer, "Human Energy Production as a Process in the Biosphere," *Scientific American*, Sept. 1970, 175–190; George M. Woodwell, "The Energy Cycle of the Biosphere," *Scientific American*, Sept. 1970, 64–74.

189. Freeman J. Dyson, "Energy and the Universe," *Scientific American*, Sept. 1971, 51.

190. Hubbert's contribution discussed the transitory nature of the fossil fuel epoch in particular. Hubbert, "The Energy Resources of the Earth," 60–70.

191. R. A. Rappaport, "The Flow of Energy in an Agricultural Society," *Scientific American*, Sept. 1971, 132.

192. The *Bulletin* was established by a group of physicists, mostly from the Manhattan Project, in December 1945 to publicize the social and political consequences of nuclear energy and to lobby for arms control. In 1947, it introduced its "Doomsday Clock" to visualize and assess the threat of nuclear war by the proximity of its time to midnight. "The Atomic Scientists of Chicago," *Bulletin of the Atomic Scientists of Chicago* 1, no.1 (Dec. 10, 1945).

193. "Foreword," *Bulletin of the Atomic Scientists* 27, no. 7 (Sept. 1971): 2.

194. Bernard I. Spinrad, "America's Energy Crisis: Reality or Hysteria," *Bulletin of the Atomic Scientists* 27, no. 7 (Sept. 1971): 3.

195. Odum, *Environment, Power, and Society*, 6

196. Odum, *Environment, Power, and Society*, 7.

197. Odum, "Energy, Ecology, and Economics," 226.

198. *Newsweek*, Jan. 22, 1973. The cover shows a light bulb, suggesting that oil had not yet become the symbol for energy shortage that it would be after the embargo and the development of oil addiction imagery. See also James E. Akins, "The Oil Crisis: This Time the Wolf Is Here," *Foreign Affairs* 51, no. 3 (April 1973): 462–490. Crucially, Akins ended the piece by arguing that the oil crisis would be short-lived because oil was unlikely to be the predominant fuel by the end of the century. Conservation and alternative energy would get the major consuming nations through the next twenty-five years. He ended with the hope that "perhaps by the year 2000" some abundant energy source, perhaps nuclear, would render all of this moot. This appeal to a vague future is a classic trope of energy crisis discourse—the idea that some alternative is inevitable, that something can be done to make more energy.

199. Nixon, "Special Message to the Congress on Energy Policy," American Presidency Project, https://www.presidency.ucsb.edu/node/255364.

200. "3 Months on the Gasoline Line," *New York Times*, April 7, 1974.

201. "The 5-Month Oil Embargo," *Washington Post*, Mar. 25, 1974.

202. These experiences were compounded by ongoing shortages of natural gas, which affected home heating and electricity availability. See Lifset, "A New Understanding."

203. "Drivers Waiting to Fill Empty Tanks Pour Out Their Many Tales of Woe," *New York Times*, Jan. 4, 1974.

204. Beale's mantra "I'm mad as hell and I'm not going to take it anymore" listed the oil crisis, alongside inflation, "the Russians," pollution, and violent crime as targets of populist anger. Huber also notes how complaints about gas lines referred to lost leisure time: Huber, *Lifeblood*, 109–110. For an example of a populist view on the issue, see "Tourism and the Energy Crisis: Need to Get Away Is Essential, Not a Frivolity," *Chicago Tribune*, Feb. 24, 1974.

205. "How the Lives of Six People Have Been Affected," *New York Times*, July 12, 1974; "The 5-Month Oil Embargo: It Brought Out the Best and Worst in the U.S.," *Washington Post*, Mar. 25, 1974.

206. "Gas Lines Stretch On and On," *Washington Post*, Feb. 19, 1974. Newspapers offered advice for how to be productive and calm while waiting for gas so as not to cut

into leisure time: "What to Do in Gasoline Line," *Boston Globe*, Feb. 15, 1974; "10 Ways to While Away Time in a Gasoline Line," *Newsday*, Feb. 13, 1974.

207. Dunaway, *Seeing Green*, 109–112.

208. Example of conspiracy theorizing: "Gas at '73 Levels: So Why All Those Lines?," *Chicago Tribune*, Jan. 31, 1974. Examples of violence: "Violence by Motorists Erupts at Gas Stations," *Washington Post*, Feb. 21, 1974; "As Fuel Gets Harder to Find, U.S. Drivers Get Angrier and Angrier," *Wall Street Journal*, Feb. 8, 1974.

209. "Pity Man, from Gas Lines Unprotected," *Newsday*, Feb. 18, 1974.

210. "Grim Thoughts from the 'Gas' Line," *New York Times*, Feb. 19, 1974. Other examples of populist response: "Energy Crisis: Drift and Confusion," *Washington Post*, Feb. 26, 1974, quotes a congressman saying, "I wonder how long it's been since President Nixon has talked with people I mean just people, not bureaucrats and politicians." "Pity Man, from Gas Lines Unprotected"; "Going into a Gas Line and Out of Our Minds," *Newsday*, Feb. 10, 1974.

211. "The State of the Real Union," *New York Times*, Feb. 2, 1974.

212. "The Way We Were: A Look Back at the Late Great Gas Shortage," *New York Times*, April 14, 1974.

213. Roitman, *Anti-Crisis*, 39.

CHAPTER 2: **"A Time to Choose"**

The main title of this chapter is taken from Energy Policy Project of the Ford Foundation, *A Time to Choose: America's Energy Future*.

Epigraphs. Snyder, "Tomorrow's Song," in *Turtle Island*, 77; Herman Kahn, "Corporate Environment Program Research Memorandum #2: Oil Prices and Energy in General," p. 7, Kahn, Herman (1), Box 9, Richard Cheney Files, GFPL.

1. Meadows et al., *The Limits to Growth*. Subsequent, post–oil embargo, reports paid more attention to energy: Mesarovic and Pestel, *Mankind at the Turning Point*, and de Montbrial, *Energy*.

2. Barney et al., *The Global 2000 Report to the President*, iii; Jimmy Carter, "The Environment Message to the Congress," American Presidency Project, https://www .presidency.ucsb.edu/node/243080.

3. Statement, "Edmund Muskie, Securing the World's Common Future," 08/25/ 1980, Global 2000[2], Box 34, Domestic Policy Staff Energy and Natural Resources Ward/Schirmer, JCPL.

4. Barney et al., *The Global 2000 Report to the President*, 1, iii–iv.

5. One million copies of *Global 2000* were distributed, and it received widespread media coverage. For more on *Global 2000*, see Box 34 in Domestic Policy Staff Energy and Natural Resources Ward/Schirmer, JCPL.

6. Simon and Kahn, *The Resourceful Earth*, 1. Think tanks such as the Heritage Foundation (est. 1973), the Manhattan Institute (est. 1978), and the Hudson Institute (est. 1961) were a key source of pro-market arguments around energy, environment, and resource issues in the 1970s. The American Enterprise Institute (est. 1938) was the most publicly active in using the energy crisis to argue for a return to markets.

7. Wohlstetter, *Herman Kahn*, 3; Kahn, *On Thermonuclear War*, 5–6, 230, 558. On

Kahn's influence in the United States, see Michael Marien, "Herman Kahn's Things to Come," *Futurist* 7, no. 1 (February 1973): 8.

8. Simon and Kahn, *The Resourceful Earth*, 3.

9. Simon and Kahn, 25.

10. Barney et al., *Global 2000*, 40. *Global 2000*'s main policy concern was population control. It recommended meeting the World Bank's goal of stabilizing the world population at ten billion as well as transitioning to alternative energy. Simon and Kahn, *The Resourceful Earth*, 7, 9, 11–12.

11. Simon and Kahn, *The Resourceful Earth*, 42.

12. Stein, *Pivotal Decade*, 1.

13. Cowie, *The Great Exception*, 9–10.

14. Walter Heller, "Kennedy's Supply-Side Economics," *Challenge* 24, no. 2 (May/June 1981): 14–18.

15. Stein, *Pivotal Decade*, 4.

16. *Time*, Dec. 31, 1965.

17. "Nixon's Program: 'I Am Now a Keynesian,'" *New York Times,* Jan. 10, 1971.

18. See Sargent, *Superpower Transformed*, for an account of limits discourse in US foreign policy.

19. Stein, *Pivotal Decade*, 14.

20. Stein, 14.

21. "Blaming America," *New York Times*, Feb. 28, 1977.

22. "The End of the Cowboy Economy," *New York Times*, Dec. 9, 1973.

23. "Is This the End of the World's Golden Age?," *Washington Post*, Apr. 24, 1977; "The Whirlwind Confronts the Skeptics," *Time*, Jan. 21, 1974, 20.

24. Stewart Udall, "The Energy Illusions," *New York Times*, Nov. 21, 1973.

25. Tom Wolfe, "The Me Decade and the Third Great Awakening," *New York Magazine*, Aug. 23, 1976, http://nymag.com/news/features/45938/.

26. "Driving toward a Crisis," *Washington Post*, Dec. 29, 1974.

27. *Time*, Dec. 31, 1973.

28. "The Painful Change to Thinking Small," *Time*, Dec. 31, 1979, 20–25.

29. Richard Nixon, "Radio Address about the National Energy Crisis," American Presidency Project, https://www.presidency.ucsb.edu/node/256566.

30. Gerald R. Ford, "Special Message to the Congress on Energy," American Presidency Project, https://www.presidency.ucsb.edu/node/257756.

31. Jimmy Carter, "Alliance to Save Energy Statement on the Formation of the Organization," American Presidency Project, https://www.presidency.ucsb.edu/node/244137.

32. Jimmy Carter, "Address to the Nation on Energy," American Presidency Project, https://www.presidency.ucsb.edu/node/243395.

33. See McAlister, *Epic Encounters*, and Zaretsky, *No Direction Home*.

34. See James E. Akins, "The Oil Crisis: This Time the Wolf is Here," *Foreign Affairs*, Apr. 1973, 462–490; Anthony Sampson, "Who's to Blame for the High Price of Oil?," *Saturday Review*, July 9–16, 1979, 64–69.

35. "The Goal Is Self-Sufficiency," *Forbes*, Dec. 1, 1973, 27.

36. "We're at War—Bloodless, Thank God," *Forbes*, Dec. 1, 1973, 19–20.

37. Letter to the editor, *Time*, Dec. 10, 1973, 2.

38. On US intervention in Mideast politics, see Westad, *The Global Cold War*; McAlister, *Epic Encounters*; Yaqub, *Imperfect Strangers*.

39. "The Energy Crunch and National Leadership," *Fortune*, Jan. 1974, 65–66; "The Energy Joyride Is Over," *Fortune*, Sept. 1972. The Ford administration shared this view. Memorandum, William Simon to Members of Committee on Energy, 3 Sep. 1974, Folder 63, Box 18, WSP, GFPL; Ronald Bass, "A National Energy Policy Philosophy," 5 June 1974, Folder 7, Box 13, WSP, GFPL.

40. Laird, *Energy—A Crisis in Public Policy*, 3. According to Laird, by 1977 the AEI had published twenty-four studies, organized four national conferences, and produced six television programs on energy policy. See American Enterprise Institute, *Government Control of Energy?*; Mitchell, *U.S. Energy Policy*; Arrow and Kalt, *Petroleum Price Regulation*; Mitchell, *Seminar on Energy Policy*; Daly, *Energy Security*.

41. Paid advertisement: "Freedom: The Reliable Energy Source," SmithKline Forum for a Healthier American Society 1, no. 7 in *Newsweek*, Nov. 19, 1979. The advertisement quotes Herbert Stein.

42. W. Philip Gramm, "The Energy Crisis in Perspective," *Wall Street Journal*, Nov. 30, 1973. Gramm went on to become a Democratic and then a Republican senator, after supporting Reagan tax cuts. In 2000, he was one of five cosponsors of the Commodity Futures Modernization Act, which kept derivatives deregulated. He has been called a "high priest of deregulation" whom some have blamed for the 2008 financial crisis. de Vogli, *Progress or Collapse*, 176.

43. "Environment in an Energy Crisis," *Newsweek*, Dec. 15, 1973, 53–54. Political scientist and principal author of the National Environmental Policy Act (1969) Lynton Caldwell argued that "an energy crisis has been built into the structure of modern industrial society. Its advent has been as predictable as the changing seasons; but the precise timing of its arrival, its incidence, and its severity have been less easily foreseen." Caldwell, "Energy and the Structure of Social Institutions," 37.

44. Hayes, preface to *Worldwatch Paper 4*, 16–17.

45. Hagen, "Teaching Ecology during the Environmental Age, 1965–1980," 704; "Dawn for the Age of Ecology," *Newsweek*, Jan. 26, 1970, 35–36.

46. Nash, *Wilderness and the American Mind*, 192; Leopold, *A Sand County Almanac*.

47. Matthiessen, *Wildlife in America*; Carson, *Silent Spring*.

48. Quoted in Egan, *Barry Commoner and the Science of Survival*, 110.

49. Dorf and Hunter, *Appropriate Visions*.

50. Ginsberg and Snyder participated with H. T. Odum in an "Energy and Consciousness" event at the University of Florida in 1975. Ginsberg subsequently described his poetry as "based on observations of the use of science and industrial damage" in America's "petrochemical wonderland." See articles from the *Independent Florida Alligator*, Box 72, MS Collection 130, Howard T. Odum Papers (HTOP), Special and Area Studies Collections, George A. Smathers Libraries (GSL), University of Florida, Gainesville, FL.

51. Murphy, *Critical Essays on Gary Snyder*. Poets Jim Harrison, Joanne Kyger, Michael McClure, and Allen Ginsberg also saw ecology as a guide to authentic being.

52. Berry, *Standing by Words*, 58ff.

53. Snyder, *Turtle Island*, 31. My interpretation draws on Yamazato, "How to Be in This Crisis: Gary Snyder's Cross-Cultural Vision in *Turtle Island*," in Murphy, *Critical Essays on Gary Snyder*, 230–231.

54. Gary Snyder interview with Michael Helm, "The Bioregional Ethic," in Snyder, *The Real Work*, 147–149.

55. Snyder, *Turtle Island*, 103. Snyder previously invoked the imagery of addiction in his 1968 poem "Oil": "the ship burns with a furnace heart / steam veins and copper nerves / quivers and slightly twists and always goes— / easy roll of hull and deep / vibration of the turbine underfoot. / bearing what all these/crazed, *hooked* nations need / steel plates and / long injections of pure oil." Snyder, *Back Country*, 26.

56. Akins, "The Oil Crisis," 462–490.

57. Richard Nixon, "Address to the Nation about Policies to Deal with the Energy Shortages, Nov. 7, 1973." Energy independence meant "meeting America's energy needs from America's own energy resources," though it did not rule out oil imports as such. Older traditions of self-reliance include aversion to involvement in European affairs and literary exhortations from Henry David Thoreau, Ralph Waldo Emerson, and Walt Whitman to be self-reliant. The realization of national—and personal—dependence on distant supplies and networks of oil was a source of deep frustration and anxiety in US political discourse.

58. "Amish Unruffled by Energy Crisis," *Christian Science Monitor*, Dec. 10, 1973, B7; Hammond, "Individual Self-Sufficiency in Energy," 278–282; "Amish Still Self-Sufficient," *Washington Post*, Feb. 2, 1977; Telleen, *The Draft Horse Primer*; "Grow Your Own Fuel," *Saturday Review*, Sept. 30, 1978, 23–28; "The Draft Horse Revival: Practical, Self-Sufficient Power for the Family Farm," *Washington Post*, Mar. 4, 1979; "The Self-Sufficient Family," *Boston Globe*, Feb. 15, 1980.

59. Zaretsky, *No Direction Home*, 80–82.

60. Horowitz, *The Anxieties of Affluence*.

61. Jimmy Carter, "Address to the Nation on Energy and National Goals: 'The Malaise Speech,'" American Presidency Project, https://www.presidency.ucsb.edu/node/249458; Lasch, *The Culture of Narcissism*; Wolfe, "The 'Me' Decade." Horowitz also notes the influence of Daniel Bell and Robert Bellah on Carter's thinking, both of whom lamented the rise of self-regard at the expense of community and the Other.

62. "Editorial Notebook: Confidence," *Washington Post*, July 21, 1979; "Speech Lifts Carter Rating to 37%," *New York Times*, July 18, 1979.

63. "Carter Finds Crisis of Confidence," *Washington Post*, July 16, 1979; "Citizens Ask if Carter Is Part of the Crisis," *New York Times*, Aug. 3, 1979; "Harris Survey: Most Have Confidence in U.S., Not in Carter," *Chicago Tribune*, Nov. 12, 1979.

64. ". . . and a Misreading of the Nation's Soul," *Wall Street Journal*, July 19, 1979; Paul W. McCracken, "Carter's Confidence Speech," *Wall Street Journal*, Aug. 1, 1979. McCracken, an economist, briefly chaired Nixon's Council of Economic Advisers as well as the AEI's Council of Economic Advisers.

65. Ronald Reagan, "Announcement for Presidential Candidacy," 11/13/1979, Miller

Center, https://millercenter.org/the-presidency/presidential-speeches/november-13 -1979-announcement-presidential-candidacy.

66. "'It's the Dumbest Experiment in Human History': Elon Musk Rails against Fossil Fuel Use and Climate Change," *Business Insider*, Sept. 8, 2018; "Divesting from Fossil Fuels," *New York Times*, Dec. 18, 2017.

67. Rickover, "Address before Minnesota State Medical Association"; Crabb, "The Energy Crisis, the Middle East, and American Foreign Policy," 48–49. Rickover used the phrase again when he testified before Congress in 1972: "Statement of Vice Adm. H.G. Rickover, U.S. Navy," in US Congress, Committee on Interior and Insular Affairs, *Fuel and Energy Resources, 1972 Part II: Hearings before the Committee on Interior and Insular Affairs*, 92nd cong., 2nd sess., 1972, 628–647. Rickover was inspired by reading Truman's *Resources for Freedom* (1952) report.

68. Rickover, "Address before Minnesota State Medical Association," 11.

69. Rickover, "Statement of Vice Adm. H.G. Rickover," 628–629.

70. Lynton Caldwell argued that the crisis had exposed the incompatibility of US socioeconomic life with the earth's energic limits. Caldwell, "Energy and the Structure of Social Institutions," 37ff. Other references to a passing "Fossil Fuel Age": "Statement of Richard M. Lahn, Sierra Club, July 10, 1973," in US Congress, Committee on Science and Astronautics, *Research, Development, and the Energy Crisis: Hearing Before Subcommittee on Energy* 93rd Cong., 1st sess., 1973, 53; Coates, Introduction to *Resettling America*, 3–4; Capra, *The Turning Point*, 30–31.

71. Rickover, "Address before Minnesota State Medical Association," 4.

72. Joint Committee on Atomic Energy, *Understanding the "National Energy Dilemma,"* 14.

73. Foley, *The Energy Question*, 21. See also Fowler, *Energy and the Environment*; Cook, *Man, Energy, Society*; Teller, *Energy from Heaven and Earth*.

74. Council on Environmental Quality and US Department of State, *Global Future: Time to Act*.

75. Several ecological economists worked for the World Bank, which produced energy conservation and development reports. Undoubtedly, the naturalization of an "Age of Scarcity" supported the bank's development initiatives in the non-Western world. World Bank, *The Energy Transition in Developing Countries*.

76. Economist Hans Landsberg argued that the energy transition would be "not a transition to a new stable condition, but to a set of new tendencies: from cheaper, abundant, reliable energy to costlier, scarcer, and more chancy energy supply . . . from taking energy for granted to paying attention to it." Landsberg, preface to Ford Foundation and Resources for the Future, *Energy*, xxii. Robert W. Tucker, "America in Decline: The Foreign Policy of 'Maturity,'" *Foreign Affairs* 58 (1979): 475; Richard L. Strout, "America's Affair with Oil," *Christian Science Monitor*, Apr. 4, 1978, 31.

77. Jon D. Roland, "Review of *World Dynamics*," *Futurist*, Aug. 1971, 145; Emile Benoit, "The Coming Age of Shortages," *Bulletin of the Atomic Scientists*, Jan. 1976, 6–16.

78. S. David Freeman of Resources for the Future characterized this period as an

"Age of Energy." S. David Freeman, "Toward a Policy of Energy Conservation," *Bulletin of the Atomic Scientists*, Oct. 1971, 8–9.

79. "Extinction for the Fossil Fuel Age?," *Washington Post*, Apr. 16, 1973.

80. Richard Eder, "Whatever Happened to the Idea of Progress?," *New York Times*, Dec. 30, 1979; Arthur Schlesinger Jr., "The End of an Era?," *Wall Street Journal*, Nov. 20, 1980; "Suburbia: End of the Golden Age," *New York Times*, Mar. 16, 1980.

81. Benoit, "The Coming Age of Shortages," 7.

82. Rufus, *Awakening from the American Dream*, 45, 89.

83. Livingston and Anderson et al., *A Desirable Energy Future*, 15–16, 42.

84. Quoted in "Suburbia: End of the Golden Age," *New York Times*, Mar. 16, 1980.

85. One sociological study funded by the National Science Foundation explored the pressures that energy scarcity put on families in areas with large suburban communities: Perlman and Warren, *Families in the Energy Crisis.*

86. Schlesinger, "The End of an Era?"

87. Mitchell, *U.S. Energy Policy*; Adelman, *The World Petroleum Market*; MacAvoy and Pindyck, *Price Controls and Natural Gas Shortage*; Adelman, *The Economics of Petroleum Supply*; MacAvoy, *Energy Policy.*

88. "The Long-Term Energy Crisis," *New York Times*, Apr. 19, 1973.

89. "The Nature of Energy," *New York Times*, Dec. 5, 1973.

90. Sabin, *Crude Politics.*

91. Huber, *Lifeblood*, 109–110, 123.

92. "Higher Gas Prices Put Crimp in California 'Cruising,' " *New York Times*, Aug. 23, 1980.

93. "High Gas Prices Force Californians to Shift Gears," *Los Angeles Times*, Apr. 16, 1980.

94. For a summary of Hubbert's studies of fossil fuel reserves, see "Statement of M. King Hubbert, U.S. Geological Survey, Department of the Interior: Methods of Estimating Oil and Gas Resources," in US Congress, Committee on Interstate and Foreign Commerce, *Energy Conservation and Oil Policy Part 1: Hearings before the Subcommittee on Energy and Power*, 94th cong., 1st sess., 1975, 671–700.

95. Hubbert, "The World's Evolving Energy System," 1007. H. T. Odum also divided history according to energy: Odum, *Environment, Power, and Society*, 230–231.

96. Lincoln Gordon, "Energy Development: Crisis and Transition," *Bulletin of the Atomic Scientists*, Apr. 1981, 24–29; Hayes, *Rays of Hope*; National Research Council Committee on Nuclear and Alternative Energy Systems, *Energy in Transition*. Even large and well-funded studies like the latter struggled to pin down the future. The CONAES report concluded: "The energy problem does not arise from an overall physical scarcity of resources. There are several plausible options for an indefinitely sustainable energy supply . . . the problem is in effecting a socially acceptable and smooth transition from gradually depleting resources of oil and natural gas to new technologies whose potentials are not now fully developed or assessed and whose costs are generally unpredictable." CONAES, xvi.

97. Energy Policy Project of the Ford Foundation, *A Time to Choose.*

98. See "The Next Crises: Population and Resources," *New York Times*, Dec. 22, 1973, and "Oil Crisis Readiness," *New York Times*, Feb. 13, 1981.

99. Daniel Yergin, "The Real Meaning of the Energy Crunch," *New York Times*, July 4, 1978.

100. The notion of a steady state economy dates back to Adam Smith and David Ricardo. American ecological economists and demographers, however, developed and popularized the concept in the mid-twentieth century with reference to population and resources. Mark W. Anderson, "Economic, Steady State," in Anderson, *Berkshire Encyclopedia of Sustainability*. For 1970s steady state theory, see Georgescu-Roegen, "Energy and Economic Myths"; Daly, *Economics, Ecology, Ethics*.

101. Commoner, *The Poverty of Power*.

102. Odum made similar arguments at the time. "The Energy Crisis: Nation Challenged," *St. Petersburg Times*, Dec. 30, 1972, Articles about Odum 1969–1983, Box 52, HTOP, GSL.

103. See Rickover, "Address before Minnesota State Medical Association."

104. On the "hard path," see Lovins, *Soft Energy Paths*, 25–28.

105. Lovins, 38–39. Lovins insisted the US hard path energy system did not use energy forms appropriate to end uses. He wanted to put human wants and needs first—for shelter, transportation, light, and food, and so forth first and then figure out how to deliver those amenities efficiently, appropriately, and elegantly, rather than bluntly using electricity or fossil fuels for everything, which he argued was wasteful. He also thought that the soft path could help to control nuclear proliferation.

106. In this, their views coincided with the "small is beautiful" movement, which advocated the downsizing of economic aspirations in the West and the adoption of small-scale technologies. The father of the movement, E. F. Schumacher, invoked the impending exhaustion of fossil fuels as one motivation for transitioning to a "Buddhist economics" that values labor as essential to human life on a scale that does not damage nature. Schumacher, *Small Is Beautiful*.

107. Lovins, *Soft Energy Paths*, 50–51.

108. Lovins, 54–55. Denis Hayes held that lower energy technologies provided "more exercise, better diets, less pollution, and other indirect benefits to human health." Hayes, *Worldwatch Paper 4*, 25ff.

109. Lovins, *Soft Energy Paths*, 151.

110. See Illich, *Tools for Conviviality*, xiii. Unlike Lovins, Illich believed that the transition to a convivial society would be painful for many people. His claim that the pain would be worse if society failed to restructure along convivial lines partook in the anticipatory rhetoric of apocalypse that characterized expert writing in the period. The last chapter of *Tools for Conviviality* begins: "If within the very near future man cannot set limits to the interference of his tools with the environment and practice effective birth control, the next generations will experience the gruesome apocalypse predicted by many ecologists. Faced with these impending disasters, society can stand in wait of survival within limits set and enforced by bureaucratic dictatorships. Or it can engage in a political process by the use of legal and political procedures." Illich, *Tools for Conviviality*, 108.

111. Illich, *Tools for Conviviality*, 28, 14. Similarly: "Recycling Social Man," *Saturday Review World*, Aug. 24, 1974, 8–10, 102–106; René Dubos, "On Growth" *New York Times*, Nov. 11, 1975; Dubos, *The Wooing of the Earth*. The 1960s produced several trenchant critiques of technology that these thinkers applied to industrial energy systems. See Ellul, *The Technological Society*, and Roszak, *The Making of a Counter Culture*.

112. Carter's *Global 2000 Report* was one of many such reports that expressed great concern about the impact of higher oil/energy prices on "least developed countries" (LDCs). See James W. Howe and the Staff of the Overseas Development Council, *Energy Policies in Developing Countries: Implications for U.S. Policy*, Folder 1266, Box 112, Series 155, RG 1.3, Collection 2F, Rockefeller Archive Center (RAC), Sleepy Hollow, NY.

113. Illich, "Energy and Social Disruption"; Robertson, "Breakdown or Breakthrough," 340–348.

114. For a succinct summary of Illich's *Energy and Equity*, see Ganapathy, "Transportation and Equity." Illich argued that the average American only achieved five miles per hour out of her automobile if one considered the number of hours pent driving, parking, repairing, and working to pay for the car.

115. Ganapathy, "Transportation and Equity," 1117.

116. An exception was Ernest Callenbach's novel *Ecotopia* (1975), which detailed an ecological utopia in the near future. In the novel, California, Oregon, and Washington secede from the United States to create a steady-state society. It is narrated by an American journalist experiencing the society for the first time, which is peaceful, equitable, and sexually liberated. Callenbach published a prequel in 1981 called *Ecotopia Emerging*.

117. Coates, *Resettling America*, 23.

118. Hayes, *Rays of Hope*, 213, 217–218.

119. Hayden, *The American Future*, 122–124. The discourse of self-sufficiency also inspired the homesteading movement. *Mother Earth News* held an annual "self-sufficiency" contest. H. T. Odum et al., "Energy Basis for the United States," June 1979," Folder Energy Basis for the United States 1 of 3, Box 51, HTO Papers, GSL.

120. Capra, *The Turning Point*, 393–396; Henderson, *Creating Alternative Futures*; Odum and Odum, *Energy Basis for Man and Nature*.

121. H. T. Odum, "An Ecological Model for the Survival Ethics of Man," Folder: An Ecological Model for the Survival Ethics of Man, Box 37, HTOP, GSL.

122. H. T. Odum and John Alexander et al., *Energy Analysis of Models of the United States: Annual Report to the Department of Energy*, Dec. 1977, Folder: Energy Analysis of Models of the United States 3 of 3, Box 51, HTOP, GSL. In this report to the DOE, Odum's team predicted that America's net energy sources would continue to decline.

123. Vacca, *The Coming Dark Age*; Ehrlich, *The End of Affluence*; Heilbroner, *An Inquiry into the Human Prospect*; Stavrianos, *The Promise of the Coming Dark Age*.

124. Ophuls, *Ecology and the Politics of Scarcity*, 225ff.

125. Cook, *Man, Energy, and Society*, 200ff, 360.

126. Odum, *Energy Basis*, 1–2.

127. Schmalz, "Energy: Today's Choices, Tomorrow's Opportunities," in Schmalz, *Energy*, 10–11; Freeman J. Dyson, "Energy in the Universe," *Scientific American*, Sept. 1971, 51.

128. Schmalz, "Energy: Today's Choices, Tomorrow's Opportunities," 3.

129. Chester Cooper and Patricia Koshel, "Energy and Development," p. 8, Folder 1222, Box 107, Series 155 Energy Policy Project Files, Ford Foundation Archives, RAC.

130. H. T. Odum, "An Ecological Model for the Survival Ethics of Man," p.1, 25 Feb. 1975, Papers/Articles: An Ecological Model for the Survival Ethics of Man 1970, 1, Box 37, HTOP, GSL; Odum, *Environment, Power, and Society*, 11.

131. McNeil, Padua, and Rangarajan, Introduction to *Environmental History as If Nature Existed*, 2.

132. Shi, *The Simple Life*, 263.

133. Shi, 264.

134. Berry, *The Gift of Good Land*, 128–130.

135. Shi, *The Simple Life*, 269.

136. "Economic Outlook," *Mother Earth News* 46 (July/Aug. 1977), 33; Bob Soleta, "Agriculture and Energy Are One and the Same," *Mother Earth News* 58 (July/Aug. 1979), 100.

137. "Energy Self-Sufficiency," *Mother Earth News* 37 (Jan. 1976), 87. "Economic Outlook," *Mother Earth News* 55 (Jan/Feb. 1979), 41–42: The author writes: "Let others—if they don't know any better—worry about shrinking dollars, rising taxes, and expanding prices. Let *their* lives rise and fall on whether or not some oil sheik wants to export petroleum. Let *them* wonder if beef prices will increase by 10 or 40 or 75 percent next year. Allow *them* to wallow in the frustrations of escalating mortgage payments and rental bills. *You'll* be too busy with your own 'problems' to notice." Those problems included whether to trade your extra eggs for a neighbor's fresh strawberries.

138. "New Year, New Plans" excerpt from John Vivians "Manual of Practical Homesteading" (1975), *Mother Earth News* 37 (Jan. 1976), 86: "Our basic goal is to become as totally self-sufficient as we can, as free from the money economy and the increasingly severe difficulties of living in a world with too many people and too few resources. Our planning envisions a future—a scenario, the futurists call it, where the United States will have run out of just about every resource but the most valuable, a proportion of the world's arable land that far exceeds our proportion of the world's population."

139. See Binkley, *Getting Loose*.

140. Gilder, *Wealth and Poverty*.

141. The futurist and early trans-humanist F. M. Esfandiary envisioned a future of "undreamed abundance. Clean, cheap, limitless energy. Limitless food. Limitless raw materials." Folder 16.3 (3 of 4): Group Press 1977–1971, Box 16, F. M. Esfandiary Papers, New York Public Library Archive (NYPLA), New York, NY.

142. "En Garde, Pessimists! Enter René Dubos," *New York Times*, Oct. 17, 1971; René Dubos, "Another Way II: Less Energy, Better Life," *New York Times*, Jan. 7, 1975. Dubos believed deeply in human ingenuity and technology to create new kinds of resources and to discover and produce more oil. René Dubos, "Energy Galore (1979)," Folder 21, Box 29, Series D851, RG450, Rockefeller University, RAC. Also see Norman Macrae, "The Coming Energy Glut," *Saturday Review*, Apr. 20, 1974; "The U.S. Is Not Drilled Out," *Wall Street Journal*, Dec. 27, 1979.

143. Dubos, "Another Way II."

144.   Richard C. Gerstenberg, "The Profit System and America's Growth," *New York Times*, Mar. 4, 1974; Glenn Hueckel, "A Historical Approach to Future Economic Growth," *Science* 187 (Mar. 14, 1975): 925–931; Stobaugh and Yergin, *Energy Future*, 8. The Harvard Business School report contrasted its support of the free market to reduce energy demand to the nostalgic pastoralism of low-energy postindustrialists. George Gilder made a case for an "entrepreneurial future" and framing notions of ecological stasis and the steady state as obstacles "to the survival of civilization" because they opposed scientific progress. Gilder, *Wealth and Poverty*, 85, 237.

145.   Dubos, "On Growth."

146.   Simon, *The Ultimate Resource*, 6.

147.   Gold, *Power from the Earth*, 145–147.

148.   Gold, *Power from the Earth*, 173; Thomas Gold and Steven Soter, "The Deep-Earth-Gas Hypothesis," *Scientific American*, June 1980, 154–161.

149.   Gold and Soter, "The Deep-Earth-Gas Hypothesis," 154.

150.   Gold, *Power from the Earth*, 174.

151.   Thomas Gold, "Rethinking the Origins of Oil and Gas," *Wall Street Journal*, June 8, 1977; "Right or Wrong, Thomas Gold Is Proving Provocative Again," *New York Times*, Oct. 7, 1977; "Astronomer Believes Oil and Gas Deposits Are Old as the Earth," *Wall Street Journal*, Dec. 13, 1983.

152.   Simon, *A Time for Action*, 46. See also Adelman et al, *No Time to Confuse*, and "Speaking of Business: Time & Energy," *Wall Street Journal*, Mar. 31, 1975.

153.   Kahn et al., *The Next 200 Years*, 1.

154.   Kahn et al., 210.

155.   "The Energy Crisis in Perspective," *Wall Street Journal*, Nov. 30, 1973.

156.   "The Energy Crisis in Perspective." Other examples of this argument include Weidenbaum, Harnish, and McGowen, *Government Credit Subsidies for Energy Development*; "Speaking of Business: Energy & Us," *Wall Street Journal*, Apr. 5, 1977. Mobil Oil produced advertisements titled "A Whale of an Oil Crisis" in the late 1970s, also arguing that free markets had solved nineteenth-century whale oil scarcity. See news clipping in Folder 49.4, Box 49, F. M. Esfandiary Papers, NYPLA.

157.   Raymond DeVoe Jr., "Up from Slavery: Every 'Energy Crisis' Has Sparked Opportunity," *Barron's National Business and Financial Weekly*, Apr. 28, 1980.

158.   Cooper, *Life as Surplus*, 10.

159.   Roitman, *Anti-Crisis*, 39, 6.

160.   Roitman, 11, 7.

161.   Roitman, 39.

162.   Koselleck, *The Practice of Conceptual History*, 241–244.

163.   Koselleck, 243.

CHAPTER 3: **"A Vibrant National Preoccupation"**

The main title of this chapter is taken from "More Bang from the BTU," *Newsweek*, Oct. 22, 1979, 87.

Epigraph. "If Supplies Fail, Try Demand," *New York Times*, October 1973.

1. Report, William K. Reilly, "The President's Statement: Conservation in the 1980s:

Building a Firm Foundation," *The Conservation Foundation: A Report for the Year 1979*, p. 4, Environmental Organizations, Box 33, Domestic Policy Staff Energy and Natural Resources Ward/Schirmer, JCPL.

2. "More Bang from the BTU," *Newsweek*, Oct. 22, 1979, 87.

3. "Sleepers Awake," *New York Times*, Sept. 30, 1974.

4. Pinchot, *The Fight for Conservation*, 44–50.

5. Scholars have yet to explain fully the popularity and historical impact of energy conservation in the 1970s. Energy history, as Brian Black notes, has focused on the technological and socioeconomic dynamics of energy production, as well as the interplay between "patterns of energy consumption" and consumer culture. Recent work explores improvements in energy efficicency, yet past efforts to reduce energy consumption have received insufficient attention. Scholars of environmental politics also tend to ignore the persistence of conservationism, preferring to trace the roots and rise of environmentalism by positing a relatively clear transition from the nationalist and utilitarian ethos of early twentieth-century conservationism to the more holistic ethos of "ecological interconnection" and protecting "the environment" in the postwar period. Black, "Energizing Environmental History," 86–89.

6. " 'Unholy Alliance' Ought to Fight Carter on Energy," *Wall Street Journal*, Aug. 31, 1979; Jacobs, *Panic at the Pump*, 8–9.

7. Ronald Reagan, "Statement on Signing Executive Order 12287, Providing for the Decontrol of Crude Oil and Refined Petroleum Products," Jan. 28, 1981, American Presidency Project, https://www.presidency.ucsb.edu/node/246753; US Department of Energy, *Securing America's Energy Future*, 1.

8. For example, see US Congress, House, Committee on Interstate and Foreign Commerce, *Energy Conservation and Oil Policy: Hearings before the Subcommittee on Energy and Power of the Committee*, 94th Cong., 1st sess., 1975.

9. Lincoln, "Energy Conservation," 161.

10. Odum, *Environment, Power, and Society*, 242, 253; "Whatever Happened to the Conservation Ethic?," *New York Times*, Jul. 31, 1975.

11. Daggett, *The Birth of Energy*, 3.

12. Marsh, *Man and Nature*, 3–5.

13. Jevons, *The Coal Question*.

14. Rabinbach, *The Human Motor*, 1–5.

15. Pinchot, *The Fight for Conservation*, 44. See also Clark, *Energy and the Federal Government*, 82–88.

16. Yergin, *The Prize*, 248–259.

17. Leopold, "The Ecological Conscience," 345; Leopold, *A Sand County Almanac*, 223–224.

18. Berry, "The Use of Energy," 85–99.

19. Daly, *Steady-State Economics*; Georgescu-Roegen, *The Entropy Law and the Economic Process*; Illich, *Energy and Equity*; Odum and Odum, *Energy Basis for Man and Nature*; Schumacher, *Small Is Beautiful*.

20. Illich, *Energy and Equity*, 17.

21. Herman Daly, introduction to *Toward a Steady-State Economy*, 14, 17, 19. The quotation is on page 17.

22. Daly, *Steady-State Economics*, 7.

23. Georgescu-Roegen, "Inequality, Limits, and Growth," 374.

24. Dunaway, "Gas Masks, Pogo, and the Ecological Indian," 78–79.

25. Odum, "The Ecosystem, Energy, and Human Values."

26. Georgescu-Roegen, "Inequality, Limits, and Growth," 361–364; 370–372, 374.

27. "Survival Theme of Fair on Coast," *New York Times*, Feb. 21, 1970.

28. "Black Environmentalists See Another Side of Pollution," *Enterprise Science Service*, Nelson Collection, Wisconsin Historical Society, http://www.nelsonearthday .net/docs/nelson_157-4_enterprise_science_black_environmentalists013.pdf.

29. Bell, *The Cultural Contradictions of Capitalism*; Lasch, *The Culture of Narcissism*.

30. Illich, *Energy and Equity*, 76.

31. Ophuls, *Ecology and the Politics of Scarcity*, 115.

32. Amory Lovins, "Energy Strategy: The Road Not Taken?," *Foreign Affairs* (Oct. 1976): 65–96.

33. Lovins, *Soft Energy Paths*, 4.

34. Talbot and Morgan, *Power and Light*.

35. See *Mother Earth News* (1970ff); Boyle, *Living on the Sun*; McClory, *How to Use Natural Energy*.

36. Eccli and Eccli, *Save Energy, Save Money*; Fritsch, *The Contrasumers*; Purcell, *The Waste Watchers*.

37. "Plowboy Interview: Gil Friend and David Morris of the Institute for Local Self-Reliance," *Mother Earth News* 36 (Nov./Dec. 1975), 6–15.

38. Friend and Morris, *Kilowatt Counter*, 28.

39. Todd and Todd, *Bioshelters, Ocean Arks, City Farming*, 1–4; "The New Alchemists," *New York Times*, Aug. 8, 1976.

40. Binkley, *Getting Loose*, 44–45. See also Trim, "A Quest for Permanence."

41. William Irwin Thompson, "Conceptualizing the Village: The Need for Villages," *Journal of the New Alchemists: Special Report: The Village as Social Ecology* no. 7 (1981): 140.

42. Robert Angevine, Earle Barnhart, and John Todd, "New Alchemy's Ark," *Journal of the New Alchemists* no. 2 (1974): 36; "New Alchemy Institute: Search for an Alternative Agriculture," *Science* 187 (Feb. 28, 1975): 729.

43. John Todd, "The Dilemma beyond Tomorrow," *Journal of the New Alchemists* no. 2 (1974): 122–128.

44. Nancy Jack Todd, "Explorations," *Journal of the New Alchemists* no. 7 (1981): 134; Todd and Todd, *Bioshelters, Ocean Arks, City Farming*, 12.

45. Gottlieb, *Forcing the Spring*; Jundt, *Greening the Red, White, and Blue*; Maher, *Nature's New Deal*; Robertson, *Malthusian Moment*; Rome, *Bulldozer in the Countryside*; Zelko, *Make It a Green Peace!* On survival, see Egan, *Barry Commoner and the Science of Survival*, 3. On "the environment," see Warde, Robin, and Sorlin, *The Environment*.

46. Yergin, *The Prize*, 537–538.

47. On petromodernity, see Szeman and Wellum, "Energy Humanities and the Petroleumscape."

48. "Research and Corruption: The Exxon-Nixxon Axis," *The Nation*, Jan. 5, 1974, 11.

49. "Fact and Comment: We are at War-Bloodless, Thank God," *Forbes*, Dec. 1, 1973, 19.

50. *Fortune*, Sept. 24, 1979; Jacobs, *Panic at the Pump*, 46–47.

51. Zaretsky, "The World as a Mirror: Narcissism, 'Malaise,' and the Middle-Class Family," in *No Direction Home*, 183–221.

52. "Presidential Message to Congress on Energy Legislation," p. 4, Folder 62 Committee on Energy: 1974 (June–July), Box 18, Subject Files (Secretary), WSP, GFPL; Gerald Ford: "Address to the Nation on Energy and Economic Programs," January 13, 1975, American Presidency Project, https://www.presidency.ucsb.edu/node/256590.

53. Jimmy Carter, "Address to the Nation on Energy and National Goals: 'The Malaise Speech,' " July 15, 1979, American Presidency Project, https://www.presidency.ucsb.edu/node/249458.

54. Central Intelligence Agency, "Political Perspectives on Key Global Issues," 3/1/1977, NLC-31-46-8-1-8, p. 1, RAC, JCPL.

55. Central Intelligence Agency, p. 3, 9.

56. Central Intelligence Agency, p. 7.

57. Undated memorandum for the president from Frank Zarb, "Proposed National Energy Goals," Energy Policy—Briefing Book for the President, Box 6, Domestic Council Michael Raoul-Duval Files, GFPL.

58. Nixon, "Special Message to the Congress Proposing Energy Emergency Legislation," 925.

59. "Decontrol of Oil—President's Statement and Fact Sheet, 7/14/75," p. 3, Statement on Decontrol of Petroleum, Box 5, Domestic Council Michael Raoul-Duval Files, GFPL.

60. Gerald Ford, "Address before a Joint Session of the Congress Reporting on the State of the Union—January 15, 1975," http://www.fordlibrarymuseum.gov/library/speeches/750028.asp.

61. Fact Sheet, S. 932: The National Energy Security Act, Energy Security Act, Box 25, Domestic Policy Staff Natural Resources Ward/Schirmer, JCPL. See Graf, "Between 'National' and 'Human' Security."

62. Brown, *Worldwatch Paper 14*, 41; Stockes, *Worldwatch Paper 17*, 42.

63. "Address by the Honorable Henry A. Kissinger to the University of Chicago, November 14, 1974," p., 22, Folder 5, Box I-27, Robert McNamara Papers, LOC.

64. "The National Energy Plan," ix–x, Energy Plan, Box 223, Council of Economic Advisers George C. Eads's Subject Files, JCPL.

65. Carter, "Address to the Nation on Energy and National Goals: 'The Malaise Speech.' "

66. Matusow, *Nixon's Economy*, 262–263; Memo, John A. Hill to Rogers Morton, Frank Zarb, et al.: Energy Chronology, March 5, 1975, Energy Chronology, Box 13, John Marsh Files, GFPL.

67. Jacobs, *Panic at the Pump*, 61–62; Walker, "The Nuclear Power Debate of the 1970s," 227.

68. Stein, *Pivotal Decade*, 209.

69. Stein, 219; "Billions Ride on Energy Corporation," *Washington Post*, Aug. 19, 1979.

70. Council on Environmental Quality, *The Good News About Energy*, 35–39.

71. Council on Environmental Quality, 47.

72. "Basis for a Federal Energy Conservation Program," p. 1, Energy-Project Independence February–July 1974(2), Box 66, Gary L. Seevers Files, Council of Economic Advisers Records 1974–1977, GFPL.

73. Energy Sense, Box 6, Frank Zarb Papers, GFPL.

74. "Ad Council Kicks Off Campaign," *Back Stage*, Jan. 4, 1974, 1.

75. Don Shula's "Don't Be Fuelish" PSA, https://www.youtube.com/watch?v=8UWi Ztrqjt8; George C. Scott's PSA, https://www.youtube.com/watch?v=5Ev9JPBd8vg.

76. Cohen, *A Consumers' Republic*, 112–165.

77. "Remarks of the President to the National Association of Counties July 16, 1979," Energy: Oil Import Quotas [1], Box 223, Staff Office—Council of Economic Advisers, George C. Eads's Subject Files, JCPL; Jimmy Carter, "Address Delivered Before a Joint Session of the Congress on the National Energy Plan," April 20, 1977, *The American Presidency Project*, https://www.presidency.ucsb.edu/node/243400.

78. "Invitation to President's Council on Energy Efficiency" and "The President's Council and Award for Energy Efficiency," in White House Office of Administration—Pickman, Energy Efficiency Program—President's Council on Energy Efficiency [2], Box 8, JCPL.

79. Memo, "for the President from Al McDonald, Anne Wexler, and Charles Duncan re: Status Report on President's Energy Conservation Initiative, July 21, 1980," Energy Conservation Outreach Program, Box 24, Domestic Policy Staff Natural Resources Ward/Schirmer, JCPL.

80. "Energy Conservation Outreach Program: Phase II: Residential-Agricultural, July 22, 1980," Energy Conservation Outreach Program, Box 24, Domestic Policy Staff Natural Resources Ward/Schirmer, JCPL.

81. The National Energy Foundation, "Energy: Its sources, its efficient utilization, its conservation" and "The Energy Crisis: Young America Faces the Challenge," Youth & Energy-National Energy Foundation [O/A 68], Box 67, Records of the Office of the Assistant to the President for Communication, JCPL.

82. General Letter from Donald D. Duggan, 02/7/1977, Youth & Energy-National Science Teachers Association-Materials on Energy [O/A 68] [1], Box 67, Records of the Office of the Assistant to the President for Communication, JCPL.

83. "NTSA: Interdisciplinary Student/Teacher Materials on Energy, The Environment, and the Economy: The Energy We Use (Grade 1)," Youth & Energy-National Science Teachers Association-Materials on Energy [O/A 68] [1], Box 67, Records of the Office of the Assistant to the President for Communication, JCPL.

84. "Young Energy Savers: Zip's Energy Club," Energy Conservation [Youth Groups & Weekly Reader] 6/77–10/77, Box 109, Office of Public Liaison Costanza, JCPL.

85. "Driving Down Gasoline Costs," *New York Times*, Jun. 28, 1980.

86. "Creating the Energy Efficient Society," *New York Times*, Sept. 23, 1979; Cambridge Reports/Research International, Survey, Apr. 1978, USCAMREP.78APR.R250 (Ithaca, NY: Cornell University, Roper Center for Public Opinion Research, iPOLL); Opinion Research Corporation, ORC Public Opinion Index, Sept. 1984, USORC,

110084, R02 (Ithaca, NY: Cornell University, Roper Center for Public Opinion Research, iPOLL).

87. ABC News / Louis Harris and Associates, Survey, Mar. 1979, USABCHS.032079. R2B (Ithaca, NY: Cornell University and Roper Center for Public Opinion Research, iPOLL); Council for Environmental Quality, Public Opinion on Environmental Issues, Jan. 1980, USRFF.80ENVR.R40 (Ithaca, NY: Cornell University and Roper Center for Public Opinion Research, iPOLL); US Energy Information Administration, *March 2019 Monthly Energy Review* (March 26, 2019), 34. https://www.eia.gov/totalenergy/data/monthly/pdf/mer.pdf.

88. "Fiddling with the Thermostat," *New York Times*, Jan. 19, 1982.

89. W. Philip Gramm, "The Energy Crisis in Perspective," *Wall Street Journal*, Nov. 30, 1973; "The 'Energy Crisis' Explained," *Wall Street Journal*, May 27, 1977; Friedman and Friedman, *Free to Choose*, 219–222.

90. US Congress, House, Committee, *Energy Conservation and Oil Policy*, 205.

91. William Simon, "The Energy Policy Calamity," *Wall Street Journal*, Jun. 10, 1977; "Government Can Help More by Doing Less," *Fortune*, Sept. 24, 1979, 85.

92. "Federal Energy Administration, 'The Effects of Decontrol, August 18, 1975,'" Decontrol of Oil—Effects of Decontrol (FEA Report), Box 4, Domestic Council Michael Raoul-Duval Files, GFPL.

93. Memo, Bill Simon to Economic Policy Group Re: Project Independence Plan and Schedule, [Undated], Energy—Project Independence, February-July 1974 (1), Box 66, Council of Economic Advisers 1974–77 Gary L. Seevers Files, GFPL.

94. "Basis for a Federal Energy Conservation Program, 3/20/74," pg. 8, Energy—Project Independence, February-July 1974 (2), Box 66, Council of Economic Advisers 1974–77 Gary L. Seevers Files, GFPL.

95. "Presidential T.V. Statement on Old Oil Decontrol," Decontrol of Oil—President's Decision, 7/10/75, Box 4, Domestic Council Michael Raoul-Duval Files, GFPL. See also "How to Achieve an American Miracle," *Business Week*, July 16, 1979; "The New Energy Paralysis," *Washington Post*, Mar. 15, 1979.

96. Conservation Foundation, *A Report*, 40.

97. Jacobs, *Panic at the Pump*, 101–113, 162–71.

98. "Letters to the Editor: Energy, the Environment and Oil-Price Controls," *Washington Post*, Mar. 24, 1979.

99. See H. T. Odum, "An Ecological Model for the Survival Ethics of Man, UNC 1970," Folder: An Ecological Model, Box 37, HTOP, GSL; "Plowboy Interview: William Ophuls," *Mother Earth News* 25 (Jan. 1974), 10; Costanza, "Embodied Energy and Economic Valuation," 1224.

100. On windfall profits and decontrol: "Letters to the Editor: Energy, the Environment and Oil-Price Controls," *Washington Post*, Mar. 24, 1979.

101. US Congress, House, Committee on Interstate and Foreign Commerce, *Middle and Long-Term Energy Policies and Alternatives: Hearings before the Subcommittee on Energy and Power*, 94th Cong., 2nd Sess., 1976, 45.

102. Landsberg et al., *Energy*, 46.

103. "Harris Survey: Public Makes Big Shift, Supports Oil Decontrol," Decontrol

of Oil—President's Veto Message September 1975, Box 5, Domestic Council Michael Raoul-Duval Files, GFPL.

104. Carter, "Address Delivered Before a Joint Session of the Congress on the National Energy Plan."

105. "The White House Fact Sheet: Energy Programs and Accomplishments," President's Energy Policy Accomplishments, Box 38, Domestic Policy Staff Energy and Natural Resources Ward / Schirmer, JCPL.

106. "Speech, Economic Growth for the 1980's," p. 6, Speeches on Economics, Box 40, Domestic Policy Staff Energy and Natural Resources Ward / Schirmer, JCPL.

107. Jimmy Carter, "Energy Address to the Nation," April 5, 1979, American Presidency Project, https://www.presidency.ucsb.edu/node/249678. Most major oil companies supported decontrol as a means to conservation and domestic production: "They Ask Decontrol Now," *New York Times*, Aug. 26, 1975; "Exxon Lauds Decontrol," *Washington Post*, Apr. 26, 1979.

108. Jimmy Carter, "Decontrol of Domestic Oil Prices Statement by the President," June 1, 1979, American Presidency Project, https://www.presidency.ucsb.edu/node /249769.

109. Ronald Reagan, "Statement on Signing Executive Order 12287," Jan. 28, 1981, American Presidency Project, https://www.presidency.ucsb.edu/node/246753.

110. US Department of Energy, *Securing America's Energy Future*, 1.

111. "The Energy Crisis Is Over!" *Harper's*, Nov. 1981, 25–36.

112. Leopold, *Sand County Almanac*, 210.

113. Berry, "The Use of Energy," 98.

CHAPTER 4: **"Put Your Foot on the Pedal"**

The main title of this chapter is taken from *White Lightning*, directed by Joseph Sargent (1973; New York, NY: Kino Lorber, 2014), DVD.

Epigraph. The Cannonball Run (dir. Hal Needham, 1981).

1. Jimmy Carter, "Address to the Nation on Energy," April 18, 1977, American Presidency Project, https://www.presidency.ucsb.edu/node/243395. Carter was alluding to William James's insight about the need to find a substitute for war to inspire national unity.

2. "A Meow in Search of an Enemy," *New York Times*, Apr. 23, 1977.

3. Huber, *Lifeblood*, xi.

4. Hamilton, *Trucking Country*, 216ff. See also "The Energy Peril to Labor Peace," *Business Week*, Dec. 8, 1973; "Truckers Protest Speed Crackdown," *Washington Post*, Nov. 4, 1975; "Truckers Block I-55 in Protest," *Chicago Tribune*, Dec. 14, 1977.

5. Anderson, "Levittown Is Burning!," 48. Protests occurred throughout the country in 1979, often involving violence: "Guard Called in Truck Strike," *Washington Post*, June 14, 1979; "14 States Affected by Trucker Protests," *New York Times*, June 8, 1979.

6. Yates, *Cannonball*, 14.

7. "*Car and Driver's* 60th Anniversary: The 1970s," *Car and Driver* (June 2015), http:// www.caranddriver.com/features/the-1970s-car-and-drivers-60th-anniversary-feature. The *Chicago Tribune* called the race a "madcap adventure sanctioned by nobody and

condemned by every law enforcement agency in the land." "This Road Race Was No Polish Joke," *Chicago Tribune*, Apr. 11, 1973. *Popular Mechanics* also framed the race as a response to the 55 mph speed limit: "The Real Gumball Rally," *Popular Mechanics* (July 1976).

8. Cover, *Car and Driver* (Aug. 1975). Five Cannonball races took place from 1971 to 1979. The rules stated that competitors could drive any vehicle at any speed following any route to achieve the fasted elapsed time from New York to Los Angeles. Brock Yates, "The Cannonball Baker Sea-to-Shining Sea Memorial Trophy Dash," *Car and Driver* (May 1972), http://www.caranddriver.com/features/the-cannonball-baker-sea-to-shining-sea -memorial-trophy-dash-archived-feature. For media coverage, see "Coast-to-Coast Auto Race Breaks Laws but Ends with Winner and Absolution," *New York Times*, Nov. 17, 1972; "Modern Living: The Cannonball Dash," *Time*, May 5, 1975; "It's Coast to Coast in Cannonball Race," *New York Times*, Apr. 2, 1979; "Heinz Car Wins Race to Coast," *New York Times*, Apr. 4, 1979.

9. As Christopher Sharrett notes, "Cinema has . . . replaced the novel in art's traditional function of illustrating the characteristics of the society in which it is produced." Sharrett, *Crisis Cinema*, 1. See also Corkin, *Starring New York*, 2–3.

10. I use the term "car film" to refer to movies that put cars, driving, and speed at the center of their plotting and aesthetics. They differ from "road movies" insofar as the latter typically use the road as a metaphor and often seek to subvert social conventions and hierarchies through narratives of self-discovery. Car films overlapped with other genres, especially westerns and comedies.

11. Laderman, *Driving Visions*, 2–3. Laderman argues that road movies celebrate "subversion as a literal venturing outside of society."

12. Harriet R. Polt, "Easy Rider," *Film Quarterly* 23, no. 1 (Autumn 1969): 23.

13. Greg Ford, "Two-Lane Blacktop," *Film Quarterly* 25, no. 2 (Winter 1971–72): 53.

14. French, " 'The Southern,' " 10–11.

15. Thomas Elsaesser, "The Pathos of Failure: American Films in the 1970s: Notes on the Unmotivated Hero," in Elsaesser, Horwath, and King, *The Last Great American Picture Show*, 280. Originally published in *Monograph* 6 (1975).

16. "Putting 'Easy Pieces' Together," *Wall Street Journal*, Oct. 5, 1970. On the working-class politics of *Five Easy Pieces*, see Nystrom, *Hard Hats, Rednecks, and Macho Men*, 39–46.

17. *Three Days of the Condor* (dir. Sydney Pollack, 1975). *Condor* was the number one film of fall 1975. "15 of 30 Films Qualify as 'Hits' During the Fall Quarter," *Box Office*, Feb. 2, 1976, 3. Alan Pakula's *Rollover* (1981) made a similar point about the political danger and corruption inherent in American dependency on Arab petrodollars. A *Boston Globe* review noted the film suggests that the United States "no longer controls its own economic and financial destiny and that this lack of control is something that evolved with very little fanfare." "Review: Rollover Pays Off," *Boston Globe*, Dec. 11, 1981.

18. Patrick McGilligan, "*Three Days of the Condor* Sidney Pollack Interviewed: Hollywood Uncovers the CIA," *Jump Cut* no. 10–11 (1976), 11–12. Quotation from Martin Knelman, "Condor: Robert Redford and the American Paranoia," *Globe and Mail*, Oct. 10, 1975.

19. Sidney Lumet's *Network* (1976) similarly warned about the moral decline of American institutions, though it was broader and more acerbic in its accusations. *Network* decried what it saw as the rampant dehumanization of American culture, which robbed individuals of their agency and eroded their trust in institutions. It blamed media and capitalist greed, though the diatribe that Paddy Chayefsky wrote for Peter Finch's iconic Howard Beale character included the energy crisis and inflation. Historians often see *Network* as representative of the mood of the 1970s. Sandbrook, *Mad As Hell.*

20. Rod MacLeish, "Hands Off 'King Kong'" *Washington Post*, Apr. 12, 1976; Jonathan Rosenbaum, "King Kong," *Monthly Film Bulletin* (Jan.1, 1977), 25–26; Vincent Canby, "'King Kong' Bigger, Not Better, in a Return to Screen of Crime," *New York Times*, Dec. 18, 1976. Pauline Kael was an exception. See "Here's to the Big One," *New Yorker*, Jan. 3, 1977.

21. Richard Eder, "Hollywood Is Having an Affair with the Anti-Hero," *New York Times*, Jan. 2, 1977.

22. Ernest Larsen, "*King Kong* meets Exxon," *Jump Cut* 16 (1977), 3–4, http://www.ejumpcut.org/archive/onlinessays/JC16folder/KingKongExxon.html. Larsen later noted that *The Towering Inferno* (1974), *Earthquake* (1974), and other disaster films shared Kong's critique of the "incompetence and/or duplicity of the managerial elite." Ernest Larsen, "*Alien. Dawn of the Dead.* Hi Tech Horror," *Jump Cut* no. 21 (Nov. 1979), 1, 12, 30. http://www.ejumpcut.org/archive/onlinessays/JC21folder/AlienDawnDead.html.

23. Pauline Kael memorably described Kong in the film as "Christ as mistreated pet." Kael, "Here's to the Big One."

24. Larsen, "King Kong Meets Exxon."

25. Richard Eder decried the equation of humanity with greed in *King Kong* and said that its message was "more on the order of a reflex than a thought." "Hollywood Is Having an Affair with the Anti-Hero."

26. Uhlman and Heitmann, "Stealing Freedom," 86.

27. Huber, *Lifeblood*, 108.

28. Landsberg et al., *Energy*; Friedman and Friedman, *Free to Choose*, 218–221.

29. Foucault, *The Birth of Biopolitics*, 215–237.

30. Binkley, *Getting Loose*, 6–16. Neal Curtis also highlights choice as the distinctive feature of neoliberal subjectivity in his critique of privatization. Curtis, *Idiotism*, 5.

31. Brown, "American Nightmare," 692.

32. Many directors of the "New Hollywood" era made at least one variation of a car or road film in the 1970s, including Steven Spielberg (*Duel* and *Sugarland Express*), Jonathan Demme (*Crazy Mama*), Peter Yates (*Bullitt; Robbery, Mother, Jugs and Speed*), David Cronenberg (*Fast Company*), Terrence Malick (*Badlands*), Bob Rafelson (*Five Easy Pieces*), Michael Cimino (*Thunderbolt and Lightfoot*), Ron Howard (*Grand Theft Auto*), and George Lucas (*American Graffiti*). More established filmmakers also tried their hand at car films, including Monte Hellman (*Two-Lane Blacktop*), William Friedkin (*Sorcerer*), Henry Fonda (*Great Smokey Roadblock*), Sam Peckinpah (*Convoy*), and Daniel Petrie (*The Betsy*). Dozens of now largely forgotten car films—sometimes deemed "carsploitation"—also proliferated. Many of these were produced by the king of low-budget schlock, Roger Corman, including *Death Race 2000, Cannonball!, Jackson County Jail, Macon County*

Line, Deathsport, Eat My Dust!, and *Fast Charlie . . . the Moonbeam Rider.* Other films concerned with automobility included *The Car, Thunder and Lightning, The Gauntlet, The California Kid, Corvette Summer, Gone in 60 Seconds, The Junkman, The Cannonball Run, The Cannonball Run II, Mad Max, Damnation Alley, White Lightning, Smash Up on Interstate 5, Stingray, Dirty Mary Crazy Larry, The Formula, The Driver,* and *Greased Lightning.* For more on road movies as a proving ground for New Hollywood directors, see Mills, *The Road Story and the Rebel,* 135.

33. Postwar films celebrating automobility included *The Fast and the Furious* (1955), *Grand Prix* (1966), *Bullitt* (1968), *Easy Rider* (1968), and *The Italian Job* (1969).

34. Nystrom, *Hard Hats, Rednecks, and Macho Men,* 56.

35. "Film Review: *American Graffiti,*" *Variety* 271, no. 6 (Jun. 20, 1973), 20.

36. Lucas has emphasized his interest in the cultural practices that accompanied youth car culture. American Film Institute, "George Lucas on *American Graffiti,*" Oct. 30, 2009. Video, 4:38. https://www.youtube.com/watch?v=WvmFpj2Bgyc.

37. Vincent Canby, "'Heavy Traffic' and 'American Graffiti'—Two of the Best," *New York Times,* Sept. 16, 1973.

38. Jameson, *The Cultural Turn,* 7–9. With the tagline "Where Were You in '62?," the advertising for *American Graffiti* was also designed to invoke nostalgia: *Independent Film Journal* 73, no. 8 (Mar. 18, 1974), 10.

39. For this line of criticism, see Sodowsky, Sodowsky, and Witte, "The Epic World of American Graffiti," and Decker, "They Want Unfreedom and One-Dimensional Thought?"

40. Sodowsky, Sodowsky, and Witte, "The Epic World of American Graffiti," 47, and Aljean Harmetz, "Our Own Past, Lost and Gone, Is in 'Graffiti,'" *New York Times,* Dec. 2, 1973.

41. Lucas quoted in Stephen Farber, "George Lucas: The Stinky Kid Hits the Big Time," *Film Quarterly* 27, no. 3 (Spring 1974), 8.

42. Mirzoeff, "Visualizing the Anthropocene," 213.

43. Gary Arnold, "Cruising through 'American Graffiti': Hot Rods, Good Times and Longings," *Washington Post,* Aug. 31, 1973.

44. Michael Dempsey, "American Graffiti," *Film Quarterly* 27, no. 1 (Autumn 1973), 58.

45. Stephen Farber, "Graffiti Ranks with Bonnie and Clyde," *New York Times,* Aug. 5, 1973; *New Yorker,* Aug. 6, 1973.

46. Dempsey, "American Graffiti," 59, notes that Lucas critiques nostalgia by showing how far away and quaint 1962 seemed to audiences in 1973. This, for Dempsey, exposes the disposability of pop culture.

47. For a mythological analysis of *American Graffiti,* see Sodowsky, Sodowsky, and Witte, "The Epic World of American Graffiti." If the film was indeed a critique of the influence of technology on American culture, it failed to critique oil.

48. Cook, *Lost Illusions,* 38.

49. Fortunately for Twentieth Century–Fox, another film on their roster picked up the slack: *Star Wars.*

50. Judith Bloch, "Damnation Alley," *Film Quarterly* 31, no. 1 (Autumn 1981): 52.

51. Brown, *Undoing the Demos,* 103.

52. Bloch, "Damnation Alley," 51–53.

53. For example, see Friedman and Friedman, *Free to Choose*, 7.

54. This subtitle references Sammy Hagar's hit song "I Can't Drive 55" (1984).

55. Stern, *Trucker*, 13; "The Trucker Mystique," *Newsweek*, Jan. 26, 1976, 44. On the political dynamics of the trucker strikes, see Jacobs, *Panic at the Pump*, 74–79.

56. This is not to be confused with efforts of oil states to secure popular support by financing public works project with oil revenues. Alizadeh, "The Political Economy of Petro Populism." For an example of petro-populist writing, see Bruce-Briggs, *The War against the Automobile*, 176–177.

57. Hamilton, *Trucking Country*, 188. Examples of trucksploitation films and television shows from this period include *White Line Fever* (1975), *Breaker Breaker!* (1977), *Smokey and the Bandit* (1977), *Smokey and the Bandit II* (1980), *Moonrunners* (1975), *Deadhead Miles* (1973), *Texas Lightning* (1981), *High Ballin'* (1978), *Steel Cowboy* (1978 TV movie), *The Great Smokey Roadblock* (1977), *Battletruck* (1982), *Truck Stop Women* (1974), *BJ and the Bear* (1978–81), and *Movin' On* (1974–76).

58. Richard Thompson, "What's Your 10–20: Redneck Movie Travel Notes," *Film Comment* 16, no. 4 (Jul/Aug. 1980), 38.

59. Whitman, "Song of the Open Road," in *Leaves of Grass*, 299.

60. Quoted in Nystrom, *Hard Hats, Rednecks, and Macho Men*, 85.

61. Mottram, "Blood on the Nash Ambassador," 110.

62. Marshall, "'We're Always Moving,'" 220.

63. Michael Bliss reads this plot point as "a symbol for the destructive potential of [Duck's] opposition to the smug liberalism of contemporary America," but this view ignores the liberalism of the truckers represented in the film. Bliss, *Justified Lives*, 293.

64. Stern, *Trucker*, 13, 36.

65. Quoted in Marshall, "'We're Always Moving,'" 214.

66. Quoted in Marshall, 213.

67. Marshall, 213.

68. Richard Combs, "Convoy," *Monthly Film Bulletin* (Autumn 1978), 257.

69. "A Hit Song Makes a Strikeout Movie," *Washington Post*, June 30, 1978.

70. Peckinpah later intimated that the emptiness of the convoy represented the emptiness of American society, but he may have been commenting out of bitterness after being excluded from the final editing process. His early enthusiasm for the film suggests that the lack of coherence in the convoy coincided with Peckinpah's Randian views about creative energy and initiative. See Marshall, "'We're Always Moving,'" and Siegel, "Passion and Poetry."

71. "Bitter Thoughts from the 'Gas' Line," *New York Times*, Feb. 19, 1974; "Higher Gas Prices Put Crimp in California 'Cruising,'" *New York Times* Aug. 23, 1980.

72. Vincent Canby, "Why Smokey and the Bandit Is Making a Killing," *New York Times*, Dec. 18, 1977.

73. Nystrom, *Hard Hats, Rednecks, and Macho Men*, 55.

74. Quoted in Nystrom, 55.

75. Nystrom, 57. Critic Richard Thompson noticed these themes at the time and located the roots of the good ole boy in Paul Newman's 1960s films. See Thompson, "What's Your 10–20?"

76. Southern states also saw film as an opportunity to change their image. Many crews on these films were non-union crews, contra traditional Hollywood practice. Julian Smith, "The Wooing of Burt Reynolds," *American Film* 5, no. 9 (July 1980): 46–50. Not everyone was happy about expansive car film shoots, however: Julian Smith, "What's in It for Us?," *American Film* 2, no. 4 (Feb. 1, 1977): 28–51.

77. Helvarg, *The War against the Greens*, 8. The wise use movement opposed environmental regulation as the intrusion of "big government." It prioritized jobs and the rights of private property owners.

78. Nystrom, *Hard Hats, Rednecks, and Macho Men*, 83.

79. A *Time* cover story on the two actors noted their similarities. "Hollywood's Honchos" *Time*, Jan. 9, 1978.

80. Gene Siskel, "Gator: Unsubtle Sequel Shows Haste Makes Waste," *Chicago Tribune*, July 20, 1976.

81. Stephen Harvey, "The Road Movie: Versatile Vehicle for a Portrait of Society," *New York Times*, June 22, 1980. Rather than disappearing, Harvey says the road movie took to "that most infinite highway of all—the cosmos" in *Star Wars* and *Star Trek*.

82. Jane Jacobs's work in Toronto and New York exemplified resistance to the metastasising of freeways. Flint, *Wrestling with Moses*.

83. Several films followed moonshiners, and trucksploitation films featured independent truckers working for themselves. These were not heroes fighting for the greater good, but rather for entrepreneurial aims.

84. "Gone in 60 Seconds Seen as Hilarious Entertainment," *Atlanta Daily World*, Mar. 13, 1975, 8. The film spawned two sequels, also made by H. B. Halicki: *The Junkman* (1982) and *Deadline Auto Theft* (1983). Its 2000 remake, starring Nicholas Cage and co-produced by Halicki's widow, earned over $100 million domestically.

85. Julian Smith, "Car Culture in the Movies: What Mad Pursuit," *American Film* 10, no. 1 (Sept. 1, 1976), 32.

86. "By Making Stars Out of Cars, a Junkman Turned Movie Mogul Has Already Grossed $10 Million," *People* (Sept. 22, 1975), https://people.com/archive/by-making -stars-out-of-cars-a-junkman-turned-movie-mogul-has-already-grossed-10-million-vol-4 -no-12/. For a more skeptical profile of Halicki, see Smith, "Car Culture in the Movies," 30–32; 49–53.

87. "By Making Stars Out of Cars."

88. A similar but less successful cult film was *The Car* (1977), which one critic dubbed "Jaws on wheels." "Feature Reviews: The Car," *Box Office* (May 16, 1977), B7. It inspired sequels and rip-offs, including *The Hearse* (1980) and John Carpenter's *Christine* (1983).

89. It has spawned three sequels. In the late 1990s, Tom Cruise considered producing and starring in a *Death Race* remake but eventually declined. The remakes that were eventually made were decidedly less high profile. The first starred Jason Statham, and its two sequels went direct to DVD.

90. Tom Shales, " 'Death Race 2000': Smashing and Ruthless Exploitation," *Washington Post*, July 11, 1976. *Rollerball* also featured an authoritarian society of the future that used blood sports to control its population.

91. Lawrence Van Gelder, "The Screen: Death Race 2000 Is Short on Satire," *New York Times*, June 6, 1975. Gene Siskel said of *Death Race 2000*, "It may be the goofiest and sleaziest film I've seen in the last five years." "Dismal Duo for a Drive-In Double Bill," *Chicago Tribune*, July 22, 1975. *Box Office* noted that *Death Race 2000* became a cult film on college campuses.

92. Such films included *Cannonball!* (1976), *The Gumball Rally* (1976), *Fast Charlie . . . The Moonbeam Rider* (1979), and *The Cannonball Run* (1981).

93. Wallace Alfred Wyss, "Shooting the Gumball Rally," *Popular Mechanics*, July 1976, 122.

94. *The Gumball Rally* (dir. Charles Bail, 1976).

95. Richard Eder, " 'Gumball Rally' Is Film That Loses Own Race," *New York Times*, Aug. 21, 1976; Julian Smith, "Hollywood Has Gone Car Crazy," *New York Times*, Oct. 10, 1976.

96. Vincent Canby, "Film: 'Cannonball Run' with Burt Reynolds," *New York Times*, June 20, 1981.

97. "Hollywood Has Gone Car Crazy," *New York Times*, Oct. 10, 1976.

98. Roger Ebert, "Journals: Roger Ebert from Chicago," *Film Comment* (Mar./Apr. 1977), 2.

99. Gene Siskel, "Dirty Mary Crazy Larry . . ." *Chicago Tribune*, July 23, 1974.

100. Gene Siskel, "Big Budget Action Films Roar, but They May Be Running Out of Gas," *Chicago Tribune*, Aug. 17, 1980.

101. Siskel, "Big Budget Action Films Roar."

102. Siskel, "Dirty Mary Crazy Larry"; Gary Arnold, "For Bored Thrillseekers," *Washington Post*, July 18, 1974.

103. Tom Buckley, "Screen: Mad Max, in Australia," *New York Times*, June 14, 1980.

104. Canby, "Why 'Smokey and the Bandit' Is Making a Killing."

105. Hitchcock, "Oil in an American Imaginary," 84.

CHAPTER 5: **"Markets Born of Shocks"**

The main title of this chapter is taken from "The Stuff of Wars: The Invasion of Kuwait Has Been the First Test for the New Oil Markets Developed in the Wake of the 1978–79 Oil Shock," *Economist*, Jan. 12, 1991, 66.

Epigraph. David S. Lewis, *Oil and Gas Investor*, 1983.

1. See Vitiello, *Trading through Time*.

2. For a discussion of some early US exchange history, see O'Sullivan, "The Expansion of the U.S Stock Market."

3. Ronald Reagan, "Statement on Signing Executive Order 12287, Providing for the Decontrol of Crude Oil and Refined Petroleum Products," Jan. 28, 1981, American Presidency Project, https://www.presidency.ucsb.edu/node/246753. Reagan's views matched the deregulatory program of the American Enterprise Institute on energy policy. Mitchell, *Perspectives on U.S. Energy Policy*.

4. US Department of Energy, *Securing America's Energy Future*, 1.

5. US Department of Energy, 19ff, 1–2.

6. US Department of Energy, 1.

7. Ronald Reagan, "Message to Congress Transmitting the National Energy Policy Plan," July 17, 1981, American Presidency Project, https://www.presidency.ucsb.edu/node/246543.

8. US Department of Energy, *Securing America's Energy Future*, 5.

9. US Department of Energy, 23–24.

10. The oil industry view was that the "so-called Wall Street refiners" [oil futures traders] "modernized the oil trade." "Rising Trades in Oil Futures Alter Pricing," *Wall Street Journal*, Feb. 29, 1984; Sluyterman, *Keeping Competitive in Turbulent Markets*, 60.

11. On the fossil economy, see Malm, "Who Lit This Fire?"

12. Hansen, "From Finance Capitalism to Financialization," 610.

13. Economists have also begun to consider the role of narratives in economic events. Shiller, "Narrative Economics."

14. Rodgers, *Age of Fracture*, 44.

15. This section draws on several books from the period, including Schwager, *Complete Guide to the Futures Markets*, 1–12; Swan, *Building the Global Market*; Pindyck, "Dynamics of Commodity Spot and Futures Markets."

16. Swan, *Building the Global Market*, 17–18.

17. Chassard and Halliwell, *NYMEX Crude Oil Futures Market*, 6; Chirichella quoted in McMahon, "Financial Settlement vs. Physical Delivery," *Futures*, June 4, 2014, http://www.futuresmag.com/2006/07/24/financial-settlement-vs-physical-delivery.

18. For example, see Peck, "Economic Role of Traditional Commodity Futures Markets," 17–19.

19. Teweles and Jones, *Futures Game*, 308–324. Teweles and Jones recount several studies from the twentieth century about who wins and who loses on futures markets. The studies show that a majority of speculators lose money, and that new and nonprofessional speculators lose a disproportionate amount of the time. While the exchanges collect handsome sums from commissions, and a few professional speculators with resources make a profit, there is a sense in which futures markets rely on a constant supply of fresh and inexperienced traders to serve as the losers who keep the market moving. Teweles and Jones (319) conclude that the odds of a speculator making a profit on the futures market in a given year are 1 in 4.

20. Markham, *History of Commodity Futures Trading*, 3.

21. Cronon, *Nature's Metropolis*, 120–124.

22. Cronon, 127–132.

23. Futures also traded successfully in St. Louis, Duluth, Kansas City, Omaha, San Francisco, Seattle, and Portland. Markham, *History of Commodity Futures Trading*, 6–8. On cotton, see Hieronymus, *Economics of Futures Trading*, 77.

24. Peck, "Economic Role of Traditional Commodity Futures Markets," 9.

25. Levy, *Freaks of Fortune*, 232.

26. Levy, 233–335; Fabian, *Card Sharps*, 153–202.

27. Levy, *Card Sharps*, 241–242, 252; Markham, *History of Commodity Futures Trading*, 9–10.

28. Legislative History of the Grain Futures Act of 1922: P.L. 67-311 (Washington, DC: Kirkland and Ellis, 1922). US Federal Legislative History Library.

29. US Commodity Futures Trading Commission, *History of the CFTC*, https://www
.cftc.gov/About/HistoryoftheCFTC/history_precftc.html.

30. Borstelmann, *The 1970s*, 54–55; Levitt, *Great Transformation to the Great Financialization*, 163. Others stress that Bretton Woods had been crumbling for years. See
Garber, "Collapse of the Bretton Woods."

31. Tamarkin, *New Gatsbys*, 76–77.

32. Melamed and Tamarkin, *Escape to the Futures*, 177. The Cato Institute repub-
lished the article that Friedman wrote for Melamed; see Friedman, "Need for Futures
Markets in Currencies."

33. CME Group, "Timeline of CME Achievements," http://www.cmegroup.com
/company/history/timeline-of-achievements.html.

34. Melamed and Tamarkin, *Escape to the Futures*, ix. As Chairman of CMEX,
Melamed vigorously pursued the expansion of futures contracts into currencies, US
Treasury Bills, and stock indexes in the 1970s and 1980s.

35. Markham, *History of Commodity Futures Trading*, 56–57.

36. Markham, 65, 71.

37. Markham, 73. For Seevers's views on futures markets with minimal regulation,
see Seevers, "Government Regulation and the Futures Markets."

38. "Worrying about Petrodollars," *Wall Street Journal*, Aug. 29, 1973.

39. Mitchell, "Fixing the Economy," 88.

40. Zaloom, *Out of the Pits*, 19.

41. "Commodity Traders Greet a New Deal in the Futures Pack," *Wall Street Journal*,
Sept. 11, 1974.

42. "Commodities Corner," *Barron's National Business and Financial Weekly*, Sept. 16,
1974.

43. For examples of these rationales, see the NYMEX ads: "You can protect yourself
against volatile oil prices!," in the *Wall Street Journal*, July 17, 1979; and "An Open Market
in Oil: To Restore Competition," in the *Washington Post*, Mar. 24, 1974.

44. Errera, "Exchanges and Their Contracts," 6; "Commodity Traders Greet a New
Deal," *Wall Street Journal*; "Oct. 23 Trading Set on Fuel Oil Futures," *New York Times*,
Sept. 27, 1974.

45. Clubley, *Trading in Oil Futures*, 36–37; Razavi and Fesharaki, *Fundamentals of
Petroleum Trading*, 7.

46. Gerald R. Ford, "Statement on the Energy Policy and Conservation Act," Dec.
22, 1975, American Presidency Project, https://www.presidency.ucsb.edu/node/257321;
"House Democrats Lost Energy Test," *New York Times*, Apr. 14, 1976; "Oil Men Want
Zarb Out," *Washington Post*, Apr. 29, 1976.

47. "Nine Are Charged with Tax Evasion in Futures Trades," *Wall Street Journal*, Apr.
14, 1978.

48. Goodman, *Asylum*, 54–56, 68–74; "Treat's Energy Expertise Is Base of Plans at
New York Mercantile Exchange," *New York Times*, May 18, 1982.

49. Errera, "Exchanges and Their Contracts," 11.

50. Goodman, *Asylum*, 87–88.

51. "Commodities: The Rising Interest in Heating Oil," *New York Times*, June 1, 1981.

See also "An Industry Icon Is Going Away: No. 2 Oil Heating Oil Contract Will Breathe Its Last," *The Barrel*, July 29, 2011, https://blogs.platts.com/2011/07/29/an_industry_ico/.

52. Errera, "Exchanges and Their Contracts," 11.

53. Martin, *Empire of Indifference*, 33.

54. Levy, *Freaks of Fortune*, 233

55. On the economic context of the 1970s, see Rodgers, *Age of Fracture*, 41–76.

56. Memo from Barry P. Bosworth to Stu Eizenstat and Jim McIntyre, "The Administration's Position on the Reauthorization of the Commodity Futures Trading Commission," Council of Economic Advisers, Charles L. Schultze, Council on Wage and Price Stability (COWPS) Files, Folder 4, COWPS 8, Box 101, JCPL.

57. COWPS press release, June 6, 1977, Council of Economic Advisers, Charles L. Schultze, COWPS Files, Folder 3, COWPS 15, Box 102, JCPL.

58. "After the Oil Glut," *New York Times*, Nov. 6, 1982.

59. "Crude Oil Futures Volume Graph 03/31/83 to 03/31/87," *Bloomberg*, Feb. 19, 2019.

60. Lower, "Regulation of Commodity Options," 1098; US Commodity Futures Trading Commission, "CFTC History in the 1980s," http://www.cftc.gov/About/History oftheCFTC/history_1980s.

61. Vitiello, *Trading through Time*, 112, 118, 134.

62. See the advertisement, "NYMEX—The Energy Exchange," in *The Economist* 299, no. 7449 (June 7, 1986): 15.

63. Vitiello, *Trading through Time*.

64. "Public Attitudes toward Investing: A Report by the New York Stock Exchange Based on a Nationwide Survey, June 1978," Office of Special Counsellor on Inflation Robert Strauss Business Community Files, folder "New York Stock Exchange—Capital Market Study [O/A 6738], Box 2, JCPL. See also Ott, *When Wall Street Met Main Street*.

65. On lobbying lawmakers, see Melamed, *Escape to the Futures*, 262–276.

66. Burns, *Treatise on Markets*, 117.

67. Hieronymus, *Economics of Futures Trading*, 71.

68. For example, see Duffie, *Futures Markets*, 3; Goss and Yamey, introduction to *The Economics of Futures Trading*, 1–2; Teweles and Jones, *Futures Game*, 6–8; Herbst, *Commodity Futures*, 2.

69. Rose, *Commodity Trading Manual*, 1–2.

70. Verleger, "Potential Impacts of Trading in Oil Futures," 120.

71. "The Futures Game: Selling the Public Short," *Washington Post*, Mar. 11, 1973.

72. Quoted in Gerald A. Pollack, "The Economic Consequences of the Energy Crisis," *Foreign Affairs* (April 1974): 452.

73. Pollack, 453.

74. New York Mercantile Exchange, *Nymex Energy Hedging Manual*, 4.

75. New York Mercantile Exchange, 5; Danielson, "Prospects for Crude-Oil Futures," 122.

76. Rosemary McFadden, preface to New York Mercantile Exchange, *NYMEX Energy Hedging Manual*.

77. Cambridge Energy Research Associates and Arthur Andersen & Co., *Future of Oil*

Prices, vi. John Treat of NYMEX predicted that this situation would persist for some time. Treat, "Future of Futures," in *Energy Futures: Trading Opportunities for the 1990s*, 335.

78. Goss and Yamey, introduction, 44–47.

79. Errera, "Exchanges and Their Contracts," 5; Clubley, *Trading in Oil Futures*, 36–37.

80. Such arguments popularized the ideas of economist Holbrook Working, one of the most influential theorists of futures trading. For more on Working's explanation of hedging and speculation, see Razavi and Fereidun, *Fundamentals of Petroleum Trading*, 75–76.

81. Prast and Lax, *Oil-Futures Markets*, 19–21; Razavi and Fesharaki, *Fundamentals of Petroleum Trading*, 71–73; Friedman, "Need for Futures Markets in Currencies," 637. One dissenting voice was Williams, *Economic Function of Futures Markets*.

82. See the NYMEX advertisement, "Options for a Volatile Market—Crude Comes of Age," *Barron's National Business and Financial Weekly*, Sept. 1, 1986.

83. Teweles and Jones, *Futures Game*, 5. See also Teweles, Harlow, and Stone, *Commodity Futures Trading Guide*, 4–6.

84. Burns, *Treatise on Markets*, 45.

85. Peck, "Economic Role of Traditional Commodity Futures Markets," in *Futures Markets*, 75.

86. Razavi and Fesharaki, *Fundamentals of Petroleum Trading*, 73–78.

87. Board of Governors of the Federal Reserve System et al., *Study of the Effects on the Economy of Trading in Futures and Options*, 11–12; see also Friedman, Harrison, and Salmon, "Informational Role of Futures Markets" in Streit, *Futures Markets*.

88. Chassard and Halliwell, *NYMEX Crude Oil Futures Market*, 9.

89. Peck, "Economic Role of Traditional Commodity Futures Markets," 74; see also Adelman, *Genie Out of the Bottle*, 193, 328.

90. Mirowski, *Never Let a Serious Crisis Go to Waste*, 435–436; Rodgers, *Age of Fracture*, 41–76.

91. Cooper, *Life as Surplus*, 43–45.

92. Hayek, "Use of Knowledge in Society."

93. Friedman, "Energy Crisis: A Humane Solution," 3, Feb. 10, 1978, Collected Works of Milton Friedman Project records, Hoover Institution Archives, https://miltonfriedman .hoover.org/objects/57283.

94. Friedman, "Energy Crisis," 3–4.

95. Friedman, "What Price Oil," *Newsweek*, Mar. 21, 1983.

96. Jickling, *Futures Markets and the Price of Oil*, 8.

97. Errera, "Exchanges and Their Contracts," 19.

98. Burns, *Treatise on Markets*, 74–75, 79.

99. Hubbert, *Nuclear Energy and the Fossil Fuels*; Hubbert, "Energy Resources of the Earth"; Meadows et al., *Limits to Growth*.

100. "The End of the Cowboy Economy," *New York Times*, Dec. 9, 1973.

101. Singer, "World Demand for Oil," in Simon and Kahn, *The Resourceful Earth*, 339.

102. Samuel Pizer, "Financial Flow Aspects of the Energy Problem," Jan. 11, 1974, Arthur Burns Papers, 1969–1978, "IMF Committee of 20: Meeting, Rome, Jan. 17–18, 1974 (2), Box B67, GFPL.

103. "The Energy Crisis in Perspective," *Wall Street Journal*, Nov. 30, 1973.

104. Friedman and Friedman, *Free to Choose*, 219; The Ford administration largely took this view of the energy crisis as well. "A National Energy Policy Philosophy," WSP, Folder 13.7, Box 13, Series IIIA, GFPL.

105. Blair, *Control of Oil*; Clubley, *Trading in Oil Futures*, 7.

106. "The World Oil Market in the Years Ahead: A Research Paper," 7/1/1979, NLC-15-90-8-15-2, p. iii, RAC, JCPL.

107. Foreign Affairs and National Defense Division et al., "Western Vulnerability," vi; Gisselquist, *Oil Prices and Trade Deficits*.

108. Foreign Affairs and National Defense et al., "Western Vulnerability," 119–122.

109. "Special Report: The Collapse of World Oil Prices," *Business Week*, Mar. 7, 1983; "The Economic Impact of Cheaper Oil," *Business Week*, Mar. 7, 1983.

110. "Energy-Guzzling: Most Consumers Are Cured," *Business Week*, Apr. 4, 1983.

111. Hirschfeld, "Fundamental Overview of the Energy Futures Markets," 75.

112. US Congress, Joint Economic Committee, *Energy in the Eighties*. Regarding the optimistic approach to oil supplies, see "Opening Statement of Senator Kennedy, Chairman," *Energy in the Eighties*, 1.

113. US Congress, Joint Economic Committee, "Statement of Thomas Gold," in *Energy in the Eighties*, 34.

114. Safer, *International Oil Policy*, 1.

115. US Congress, Joint Economic Committee, *Energy in the Eighties*, 8–11.

116. Safer, *International Oil Policy*, 57–61.

117. Safer, 87.

118. Safer, 121.

119. Walter J. Levy, "Oil and the Decline of the West," *Foreign Affairs* 58, no. 5 (Summer 1980): 1001.

120. "Wresting Control from OPEC: Futures Trading on the New York Mercantile Exchange Has Rewritten the Rules on Oil Pricing," *Washington Post*, Oct. 29, 1989.

121. "Why We Must Act Now," *Newsweek*, July 16, 1979; Pollack, "Economic Consequences of the Energy Crisis," 469.

122. William M. Brown and Herman Kahn, "Why OPEC Is Vulnerable," *Fortune*, July 14, 1980, 67–69.

123. "Busy Commodity Exchanges Eye Expansion of Oil Futures Trading," *Oil and Gas Journal*, Oct. 5, 1981, 49.

124. John Morris, "Energy Futures vs Cartel: Counterpoints in Oil Prices," *American Banker*, July 30, 1985, 19; see also "Wresting Control from OPEC," *Washington Post*.

125. " 'Open Outcry,' Chaos Part of Trading Art," *Chicago Tribune*, June 6, 1988. This statement was made in defense of open outcry after it came under scrutiny in the wake of the 1987 crash. Another work on oil futures trading called it "the best example of the marketplace in action." Prast and Lax, 6.

126. Yergin, *The Prize*, 707–708.

127. For discussions of how this dynamic worked, see "Trading Crude Oil Futures," *New York Times*, Mar. 28, 1983; "Rising Trades in Oil Futures Alter Pricing," *Wall Street Journal*; "Wresting Control from OPEC," *Washington Post*; Yergin, *The Prize*, 706–708.

128. Yergin, *The Prize*, 702.

129. "Rising Trades in Oil Futures Alter Pricing," *Wall Street Journal*.

130. Yergin, *The Prize*, 708.

131. "Setting Crude Prices in the Pits," *New York Times*, Dec. 9, 1984.

132. "Oil Pits Move to Center Stage," *New York Times*, Jan. 10, 1985.

133. "Oil Futures: A Gusher—or a Dry Hole?," *Newsweek*, Apr. 11, 1983.

134. "Controlled Frenzy in the Pits," *New York Times*, Aug. 1, 1981.

135. "Oil Pits Move to Center Stage," *New York Times*, Jan. 10, 1985.

136. "Oil Pits Move to Center Stage."

137. "An Uncertain and Rosy Future for the Traders on NYMEX," *Guardian*, Apr. 27, 1987.

138. "Fuel Producers Cut Exploration in U.S.," *New York Times*, Jan. 6, 1986; "Oil and Gas Industry Continues to Contract," *New York Times*, Sep. 27, 1989.

139. On concern about overconsumption in the 1970s, see Horowitz, *Anxieties of Affluence*, 203–224.

140. "An Uncertain and Rosy Future."

141. "The Oil Bust Panic," *New Republic*, Feb. 21, 1983, 16.

142. "An Uncertain and Rosy Future."

143. "In Chicago, the Future Is Now," *New York Times*, Feb. 19, 1984.

144. "An Uncertain and Rosy Future."

145. "Futures Are Looking Up," *Newsweek*, Nov. 30, 1981.

146. "Playing the Futures Game," *Newsweek*, Jan.15, 1979.

147. Oil's downward slide generated uncertainty about where oil prices would eventually settle. "The Big Oil Spill: Is It Bullish or Bearish?," *Barron's National Business and Financial Weekly*, Jan. 27, 1986; Gately, "Lessons from the 1986 Oil Price Collapse," 284.

148. "Setting Crude Prices in the Pits."

149. "OPEC, Meet Adam Smith," *Newsweek*, Oct. 29, 1984.

150. Quoted in "Futures Exchanges Affirmed as Best Method for Trading, by Study," Associated Press, Oct. 1, 1986; "Futures Market: New Dimensions in Crude Oil, Products Trading," *Oil and Gas Journal*, Mar. 5, 1984, 41.

151. "The Unrigging of Oil Prices," *Newsweek*, Mar. 7, 1983. This article was not about the effect of oil futures trading as it was published weeks before the crude contract began trading. Rather, it was about OPEC's internal struggles. It expresses the sense of victimization at the hands of OPEC that had emerged in US popular discourse, and the desire for revenge through lower oil prices.

152. "Poll: Americans and Their Money," *Money Magazine*, May 1986, Roper Center for Public Opinion Research, Cornell University, https://ropercenter.cornell.edu/.

153. Zaretsky, *No Direction Home*.

154. Ott, *When Main Street Met Wall Street*; Hyman, *Debtor Nation*.

155. Hansen, "From Finance Capitalism to Financialization," 630.

156. Beckert, *Imagined Futures*, 1–2. On narrative aspects of financialization, see Haiven, *Cultures of Financialization*; Ho, *Liquidated*; Martin, *Financialization of Daily Life*.

EPILOGUE: **Enduring Crisis**

1. Massumi, "The Future Birth of the Affective Fact."

2. "Widening Gulf: Attack on Saudi Tanker Near Oil Fields Sparks Fears of a Broader War," *Wall Street Journal*, May 17, 1984.

3. "Energy Prices Soar on Report of Attack on Persian Gulf Tanker," *Wall Street Journal*, May 17, 1984.

4. "Futures/Options: Crude Oil Futures at a 9-Week High," *New York Times*, May 3, 1986. The Soviet shutdown was expected to boost Soviet oil demand temporarily. Maurice Adelman criticized the market for moving in response to a rumor that would have no real effect on the world oil market. Regarding Mideast tensions: "Oil Prices Rise Amid Reports Iraq Jets Attacked Operations at Iran Terminal," *Wall Street Journal*, Aug. 13, 1986; "Oil Prices Climb to Nearly $20 a Barrel on Mideast Tension, Then Retreat a Bit," *Wall Street Journal*, Sept. 1, 1987.

5. Quote from "Crude-Oil Futures Rocket in Last-Minute Rally as OPEC Ministers Prepare to Meet in Madrid," *Wall Street Journal*, Oct. 20, 1988. See also "Crude-Oil Prices Soar to 4-Month Highs as OPEC Plans to Meet Other Producers," *Wall Street Journal*, Apr. 12, 1988. API figures were also said to shape markets despite the fact that they would "likely will be revised—possibly by a large margin—in the next week." "API Oil Figures Sway Futures Markets whether They Are on the Mark or Not," *Wall Street Journal*, Dec. 31, 1987.

6. Parra, *Oil Politics*, 293ff; Yergin, *The Prize*, 770–772.

7. Goldstein, "1990," 687. See also Edward L. Morse, "The Coming Oil Revolution," *Foreign Affairs* (Winter 1990/1991), 36–56. Morse was a publisher of *Petroleum Intelligence Weekly* and served Jimmy Carter as Deputy Assistant Secretary of State for International Energy Policy from 1979 to 1981.

8. "America's Neck and a New Noose," *New York Times*, Aug. 2, 1990; "Oil-Price Climb Slows; New Stability Possible," *New York Times*, Aug. 8, 1990.

9. Yergin, *The Prize*, 774–777.

10. "Brace Yourself for a Gouging," *Star-Tribune Newspaper of the Twin Cities Mpls.–St. Paul*, Aug. 7, 1990.

11. "Oil Traders Taste Blood as Futures Prices Soar," *The Times*, Aug. 7, 1990.

12. Goldstein, "1990," 687; "Is a Recession Now Inevitable? European, Japanese Booms May Cushion Third Oil Shock," *Boston Globe*, Aug. 7, 1990; "Looking Ahead, Guided by the Past," *New York Times*, Dec. 28, 1990; "Oil Remains Driving Force of the Persian Gulf War," *Washington Post*, Jan. 23, 1991.

13. US Congress, Senate, Committee on Governmental Affairs, *Oil Prices and Supplies in the Wake of the Persian Gulf Crisis: Hearings before the Committee on Governmental Affairs*, 101st Cong., 2nd Sess., Oct. 25, 31, Nov. 1, 1990, 364.

14. "Gulf Crisis to Keep Oil Markets in Spotlight," Reuters News, Aug. 10, 1990.

15. "So Why Are Oil Prices So High?," *Newsweek*, Oct. 8, 1990.

16. "Gulf Crisis Keeps NYMEX Traders in the Dark," *Star Tribune Newspaper of the Twin Cities Mpls.–St. Paul*, Oct. 28, 1990.

17. "Crude Oil Futures End Mixed; Crisis Keeps Markets Nervous," *Journal Record*, Aug. 16, 1990.

18. "Crude Oil, Gold Tumble on Perceived Ease in Gulf Tensions," Reuters News, Oct. 2, 1990.

19. "Crisis in the Gulf: Iraq's Invasion of Kuwait," *Washington Post*, Aug. 8, 1990; "Oil Prices Rise on Bush's Warning to Saddam," *Reuters*, Nov. 22, 1990.

20. "Price of Oil Plunges Record $5.51 A Barrel," *Journal Record*, Oct. 23, 1990.

21. "Traders in the Slippery Oil Market Bet That Recent Slide Won't Last," *Wall Street Journal*, Oct. 25, 1990.

22. "U.S. Oil Prices Volatile in Rumor Driven Market," Reuters, Aug. 9, 1990.

23. Tony Lentine, VP of Mitchell Energy: "Every time Saddam Hussein sneezes, the futures market reacts," quoted in "So Why Are Oil Prices So High?"

24. "Oil Price Volatility Persists on Diet of Rumors," Reuters, Sept. 7, 1990.

25. "In the Oil Trading Pit, Reflexes Must Be Fast," *New York Times*, Sept. 26, 1990.

26. "My Life Dealing Oil in the Pits: NYMEX's Traders Attuned to Money, Not Morality," *Washington Post*, Aug. 26, 1990.

27. "In the Oil Trading Pit, Reflexes Must Be Fast," 1.

28. US Congress, Senate, Committee, *Oil Prices and Supplies in the Wake of the Persian Gulf Crisis*, 3.

29. US Congress, Senate, Committee, *Oil Prices and Supplies in the Wake of the Persian Gulf Crisis*, 5.

30. US Congress, Senate, Committee, *Oil Prices and Supplies in the Wake of the Persian Gulf Crisis*, 336.

31. US Congress, Senate, Committee, *Oil Prices and Supplies in the Wake of the Persian Gulf Crisis*, 353.

32. "Energy Futures Faulted as Oil Prices Surge," Reuters, Sept. 28, 1990. Another analyst told Bush to "get an economics textbook." See also "Energy Futures: The New Barometer," *Energy User News*, Sept. 1, 1990, 22.

33. "Oil-Price Role Defended by Futures Exchanges," *New York Times*, Nov. 12, 1990.

34. Keynes, *The General Theory*, 161–162. Keynes famously described the irrational side of human economic decisions: "Most, probably, of our decisions to do something positive, the full consequences of which will be drawn out over many days to come, can only be taken as a result of animal spirits—of a spontaneous urge to action rather than inaction, and not as the outcome of a weighted average of quantitative benefits multiplied by quantitative probabilities."

35. Roberts, *The End of Oil*; McKillop and Newman, *The Final Energy Crisis*; Tertzakian, *A Thousand Barrels a Second*; Heinberg, *Peak Everything*; Deffeyes, *Hubbert's Peak*; Partanen, Paloheimo, and Waris, *The World after Cheap Oil*. Several peak oil and anti-oil documentaries also appeared in this period, including *The End of Suburbia* (2004); *Crude Awakening: The Oil Crash* (2006); *Escape from Suburbia: Beyond the American Dream* (2007); *Crude Impact* (2006); *Gashole* (2010). For critical analysis of the peak oil movement, see Schneider-Mayerson, *Peak Oil*. Experts still talk about how to end the "fossil fuel era": Princen, Manno, and Martin, *Ending the Fossil Fuel Era*.

36. McQuaig, *It's the Crude, Dude*; Leech, *Crude Interventions*; Sandalow, *Freedom from Oil*; Nikiforuk, *The Energy of Slaves*; Coll, *Private Empire*.

37. "Bush: 'America Is Addicted to Oil,'" *New York Times*, Feb. 1, 2006. Barack Obama also decried America's oil addiction after the 2010 Deepwater Horizon spill in the Gulf of Mexico: http://www.npr.org/templates/story/story.php?storyId=127862326.

38. Clarke, *The Battle for Barrels*; Mills, *The Myth of the Oil Crisis*; Shiva, *Soil Not Oil*; Lacalle and Parrilla, *The Energy World Is Flat*.

39. On the "shale revolution," see Zuckerman, *The Frackers*; Gold, *The Boom*; Aguilera and Radetzki, *The Price of Oil*.

40. Energy Information Administration Crude Oil Futures Contract 1, https://www.eia.gov/dnav/pet/hist/LeafHandler.ashx?n=PET&s=RCLC1&f=D.

41. US Congress, Senate, Committee on Homeland Security and Governmental Affairs and Committee on Energy and Natural Resources, *Speculation in the Crude Oil Market: Joint Hearing before the Permanent Subcommittee on Investigations and the Subcommittee on Energy*, 110th Cong., 1st Sess., Dec. 11, 2007; US Congress, Senate, Committee on Homeland Security and Governmental Affairs, *The Role of Market Speculation in Rising Oil and Gas Prices: A Need to Put the Cop Back on the Beat: Staff Report by the Permanent Subcommittee on Investigations*, 109th Cong., 2nd Sess., June 27, 2006; Testimony of Michael W. Masters before the Committee on Homeland Security and Governmental Affairs United States Senate (May 20, 2008), https://www.hsgac.senate.gov/imo/media/doc/052008Masters.pdf?attempt=2.

42. US Congress, Senate, Committee, *Speculation in the Crude Oil Market*, 32–59.

43. Labban, "Oil in Parallax," 550.

44. According to the Energy Information Administration, total US energy consumption has continued to rise, https://www.eia.gov/totalenergy/data/browser/.

45. Clayton, *Market Madness*, vii–xviii, 177.

46. Roitman, *Anti-Crisis*, 2.

47. Starn, "Historians and 'Crisis'"; Shank, "Crisis"; Kuhn, *The Structure of Scientific Revolutions*; Golinksi, *Making Natural Knowledge*.

48. Morton, *Hyperobjects*.

Bibliography

Archival Collections

Jimmy Carter Presidential Library and Museum (JCPL), Atlanta, GA
    Records of the National Security Advisor via CIA Records Search Tool (CREAST)
    Staff Office Files
        Assistant for Communications
        Assistant for Public Liaison
        Council of Economic Advisers
        Domestic Policy Staff
        Special Counsellor on Inflation
        White House Office of Administration
        White House Press Office
Gerald R. Ford Presidential Library and Museum (GFPL), Ann Arbor, MI
    Richard B. Cheney Files, 1974–77
    Council of Economic Advisers Records, 1974–77
    John Marsh Files, 1974–77
    National Security Advisor's Files, Memoranda of Conversations, 1973–77
    Michael Raoul-Duval Papers, 1974–77
    William E. Simon Papers, 1972–77 (WSP)
    Frank Zarb Papers, 1974–78
Library of Congress, Manuscript Division (LOC), Washington, DC
    Robert McNamara Papers
National Archives and Records Administration, College Park, MD
    Records of the Department of the Treasury
National Museum of American History, Archives Center (NMAH), Washington, DC
    American Petroleum Institute Photograph and Film Collection, 1860s–1990
New York Public Library Archives (NYPLA), New York, NY
    F.M. Esfandiary Papers
The Rockefeller Archive Center (RAC), Sleepy Hollow, NY
    Ford Foundation Archives
    René Dubos Papers
    Rockefeller Foundation Archives
Special and Area Studies Collections, George A. Smathers Libraries, University of
    Florida (UFL), Gainesville, FL
    Howard T. Odum Papers (HTOP)

Online Archival Collections

The American Presidency Project
    Presidency Documents Archive, https://www.presidency.ucsb.edu/documents
Bloomberg Terminal, Milt Harris Library, Rotman School of Management, University of
    Toronto, Toronto, ON, Canada
Hoover Institution Archives, https://miltonfriedman.hoover.org/collections
Roper Center for Public Opinion Research, iPoll, Cornell University, Ithaca, NY, https://
    ropercenter.cornell.edu/ipoll/
US Energy Information Administration, https://www.eia.gov/
US Federal Legislative History Library, HeinOnline, http://eresources.loc.gov/record
    =e1000050~S9

Primary Sources
Periodicals

American Banker
American Film
Atlanta Daily World
The Atlantic
Barron's National Business and Financial
    Weekly
Boston Globe
Box Office
Broadcasting
Bulletin of the Atomic Scientists
Business Week
Car and Driver
Chicago Tribune
Christian Science Monitor
Department of State Bulletin
The Economist
Energy User News
Film Comment
Film Quarterly
Forbes
Foreign Affairs
Fortune
The Futurist
Globe and Mail
The Guardian
Harper's Magazine
Independent Film Journal

Journal of the New Alchemists
Journal Record
Jump Cut
Life
Mother Earth News
Monthly Film Bulletin
The Nation
National Review
New Republic
Newsday
Newsweek
New Yorker
New York Magazine
New York Times
Oil and Gas Journal
Platt's Oilgram News
Popular Mechanics
Reuters
Saturday Review
Science
Scientific American
Star-Tribune Newspaper of the Twin Cities
        Mpls.–St. Paul
Time
Variety
Wall Street Journal
Washington Post

Books, Reports, and Journal Articles

Adelman, Morris. *The Economics of Petroleum Supply: Papers by M. A. Adelman, 1962–1993.* Cambridge, MA: MIT Press, 1993.

Adelman, Morris. *The World Petroleum Market.* Baltimore: Johns Hopkins University Press, 1972.

Adelman, Morris, et al. *No Time to Confuse: A Critique of the Final Report of the Energy Policy Project of the Ford Foundation, A Time to Choose America's Energy Future.* San Francisco: Institute for Contemporary Studies, 1975.

American Enterprise Institute. *Government Control of Energy?* Washington, DC: American Enterprise Institute, 1973.

Arrow, Kenneth J., and Joseph P. Kalt. *Petroleum Price Regulation: Should We Decontrol?* Washington, DC: American Enterprise Institute, 1979.

Barney, Gerald O. *The Global 2000 Report to the President—Entering the Twenty-First Century: A Report.* Washington, DC: US Government Printing Office, 1981.

Baudrillard, Jean. *America.* London: Verso, 2010.

Bell, Daniel. *The Cultural Contradictions of Capitalism.* New York: Basic Books, 1975.

Berry, Wendell. *The Gift of Good Land: Further Essays Cultural and Agricultural.* Berkeley: Counterpoint, 2015. First published 1981.

Berry, Wendell. *Standing by Words.* San Francisco: North Points Press, 1983.

Berry, Wendell. "The Use of Energy." In *The Unsettling of America: Culture and Agriculture.* Berkeley, CA: Counterpoint Press, 2015, 85–99. First published 1977.

Blair, John. *The Control of Oil.* New York: Pantheon Books, 1976.

Boulding, Kenneth. Foreword to *Population: The First Essay,* by Thomas Robert Malthus. Ann Arbor: University of Michigan Press, 1959.

Boyle, Godfrey. *Living on the Sun: Harnessing Renewable Energy for an Equitable Society.* London: Calder and Boyars, 1975.

Brown, Harrison, James Bonner, and John Weir. *The Next Hundred Years: A Discussion Prepared for Leaders of American Industry.* New York: Viking Press, 1963.

Brown, Lester R. *Worldwatch Paper 14: Redefining National Security.* Washington, DC: Worldwatch Institute, 1977.

Bruce-Briggs, B. *The War against the Automobile.* New York: E. P. Dutton, 1975.

Burns, Arthur M. *A Treatise on Markets: Spots, Futures, Options.* Washington, DC: American Enterprise Institute, 1979.

Caldwell, Lynton. "Energy and the Structure of Social Institutions." *Human Ecology* 4 (January 1976): 31–45.

Callenbach, Ernest. *Ecotopia: The Notebooks and Reports of William Weston.* Berkeley, CA: Banyan Tree Books, 1975.

Cambridge Energy Research Associates and Arthur Andersen & Co. *The Future of Oil Prices: The Perils of Prophecy.* Cambridge, MA: Cambridge Energy Associates, 1984.

Capra, Fritjof. *The Turning Point: Science, Society, and the Rising Culture.* New York: Simon and Schuster, 1982.

Carson, Rachel. *Silent Spring.* Boston: Houghton, 1962.

Carter, Susan B., Scott Sigmund Gartner, Michael R. Haines, Alan L. Olmstead, Richard

Sutch, and Gavin Wright. *Historical Statistics of the United States: Earliest Times to the Present, Millennial Edition, Volume 4, Part D: Economic Sectors*. New York: Cambridge University Press, 2006.

Chassard, Christophe. *Option Trading and Oil Futures Markets*. Oxford: Oxford Institute for Energy Studies, 1987.

Chassard, Christophe, and Mark Halliwell. *The NYMEX Crude Oil Futures Market: An Analysis of Its Performance*. Oxford: Dot Press, 1986.

Clubley, Sally. *Trading in Oil Futures*. New York: Nichols Publishing Company, 1986.

Coates, Gary, ed. *Resettling America: Energy, Ecology, and Community*. Andover, MA: Brick House Publishing, 1981.

Commoner, Barry. *The Poverty of Power: Energy and the Economic Crisis*. New York: Knopf, 1976.

Cook, Earl. *Man, Energy, Society*. San Francisco: W. H. Freeman, 1976.

Costanza, Robert. "Embodied Energy and Economic Valuation." *Science* 210 (Dec. 12, 1980): 1224.

Cottrell, Fred. *Energy and Society: The Relation between Energy, Social Change, and Economic Development*. New York: McGraw-Hill Books, 1955.

Council on Environmental Quality. *The Good News About Energy*. Washington, DC: Government Printing Office, 1979.

Council on Environmental Quality and US Department of State. *Global Future: Time to Act*. Washington, Government Printing Office: January 1981.

Daly, Herman E., ed. *Economics, Ecology, Ethics: Essays toward a Steady State Economy*. San Francisco: W. H. Freeman and Company, 1980.

Daly, Herman E. *Steady-State Economics: The Economics of Biophysical Equilibrium and Moral Growth*. San Francisco: W. H. Freeman and Company, 1977.

Daly, Herman E. *Toward a Steady-State Economy*. San Francisco: W. H. Freeman and Company, 1973.

Daly, John Charles, mod. *Energy Security: Can We Cope with a Crisis?* Washington, DC: American Enterprise Institute, 1981.

Danielson, Albert L. "Prospects for Crude-Oil Futures." In *Oil-Futures Markets: An Introduction*, edited by William G. Prast and Howard L. Lax, 113–124. Lexington, MA: Lexington Books, 1983.

Darmstadter, Joel, with Perry D. Teitelbaum and Jaroslave G. Polach. *Energy in the World Economy: A Statistical Review of Trends in Output, Trade, and Consumption since 1925*. Baltimore: Johns Hopkins Press for Resources for the Future, 1971.

de Montbrial, Thierry. *Energy: The Countdown*. New York: Pergamon Press, 1979.

Dorf, Richard C., and Yvonne L. Hunter, eds. *Appropriate Visions: Technology, Environment and the Individual*. San Francisco: Boyd and Fraser, 1978.

Dubos, René. *The Wooing of the Earth*. New York: Scribner, 1980.

Duffie, Darrell. *Futures Markets*. Englewood Cliffs, NJ: Prentice Hall, 1989.

Dupree, Walter G., and James A. West. *United States Energy through the Year 2000*. Washington, DC: Department of the Interior, 1972.

Eccli, Eugene, and Sandra Eccli. *Save Energy, Save Money*. Washington DC: National Center for Community Action Network Services: Energy, 1974.

Egerton, Sir A. C. "Energy in the Service of Man." Paper No. 1, UNESCO Discussion Theme 1951. Paris: UNESCO, 13 February 1951. http://unesdoc.unesco.org/images /0003/000307/030755EB.pdf.

Ehrlich, Paul. *The Population Bomb*. New York: Ballantine Books, 1968.

Ehrlich, Paul, and Ehrlich, Anne. *The End of Affluence: A Blueprint for Your Future*. New York: Ballantine Books, 1974.

Ellul, Jacques. *The Technological Society*. Translated by John Wilkinson. New York: Knopf, 1964.

Energy Policy Project of the Ford Foundation. *A Time to Choose: America's Energy Future*. Cambridge, MA: Ballinger, 1974.

Energy Resources Committee. *Energy Resources and National Policy*. Washington, DC: Government Printing Office, 1939.

Errera, Steven. "Exchanges and Their Contracts." In *Energy Futures: Trading Opportunities for the 1980s*, edited by John Elting Treat, 3–21. Tulsa, OK: PennWell Books, 1984.

Federal Energy Administration. *Project Independence: Federal Energy Administration Blueprint Transcript of Fourth Public Hearing, Seattle, Washington, Sept. 5–7, 1974*. Washington, DC: Federal Energy Administration, 1974.

Foley, Gerald. *The Energy Question*. London: Penguin Books, 1976.

Foreign Affairs and National Defense Division, Economic Division, Environment and Natural Resources Policy Division. "Western Vulnerability to a Disruption of Persian Gulf Oil Supplies: U.S. Interests and Options." *Congressional Research Service Report 83–24F*. March 24, 1983.

Fowler, John. *Energy and the Environment*. New York: McGraw-Hill, 1975.

French, Warren. "'The Southern': Another Lost Cause?" In *The South and Film*, edited by Warren French. Jackson: University Press of Mississippi, 1981.

Fried, Edward R. "After the Oil Glut." *Brookings Bulletin* 18, no. 3/4 (Winter–Spring 1982): 6–11.

Friedman, Milton. "The Energy Crisis: A Humane Solution." In *The Economics of Freedom*. Cleveland: Standard Oil Company of Ohio, 1978.

Friedman, Milton. "The Need for Futures Markets in Currencies," *Cato Journal* 31, no. 3 (Fall 2011): 635–641.

Friedman, Milton, and Rose Friedman. *Free to Choose: A Personal Statement*. New York: Harcourt Brace Jovanovich, 1980.

Friend, Gil, and David Morris. *Kilowatt Counter*. Milaca, MN: Alternative Sources of Energy Magazine, 1974.

Fritsch, Albert J. *The Contrasumers: A Citizen's Guide to Resource Conservation*. New York: Praeger, 1974.

Gallacher, William. *Winner Takes All: A Privateer's Guide to Commodity Trading*. Toronto: Midway Publications, 1983.

Garvey, Gerald. *Energy, Ecology, Economy*. New York: W. W. Norton, 1972.

Gately, Dermot. "Lessons from the 1986 Oil Price Collapse." *Brookings Papers on Economic Activity* 17, no. 2 (1986): 217–284.

Gaucher, L. P. "Energy Sources of the Future for the United States." *Solar Energy* 9, no. 3 (July–September 1965): 119–126.

Georgescu-Roegen, Nicholas. "Energy and Economic Myths." *Southern Economic Journal* 41 (January 1975): 347–381.

Georgescu-Roegen, Nicholas. *The Entropy Law and the Economic Process.* Cambridge, MA: Harvard University Press, 1971.

Georgescu-Roegen, Nicholas. "Inequality, Limits, and Growth from a Bioeconomic Viewpoint." *Review of Social Economy* 35, no. 3 (December 1977): 361–375.

Gilder, George. *Wealth and Poverty.* New York: Basic Books, 1980.

Gisselquist, David. *Oil Prices and Trade Deficits: U.S. Conflicts with Japan and West Germany.* New York: Praeger Publishers, 1979.

Gold, Thomas. *Power from the Earth: Deep Earth Gas Energy for the Future.* London: Dent, 1987.

Gold, Thomas, and Steven Soter. "Abiogenic Methane and the Origin of Petroleum." *Energy Exploration and Exploitation* 1, no. 2 (1982): 89–104.

Goldstein, Walter. "1990: Is This the Third Oil Shock?" *Energy Policy* (October 1990): 686–688.

Goss, B. A., and B. S. Yamey, eds. *The Economics of Futures Trading.* New York: Macmillan, 1976.

Hammond, Allen L. "Individual Self-Sufficiency in Energy." *Science* 184 (April 19, 1974): 278–282.

Hardin, Garrett. "The Tragedy of the Commons." *Science* 162 (December 13, 1968): 1243.

Harrison, Gordon. *Earthkeeping: The War with Nature and a Proposal for Peace.* Boston: Houghton Mifflin, 1971.

Hayden, Tom. *The American Future: New Visions Beyond Old Frontiers.* Boston: South End Press, 1980.

Hayek, Friedrich. "The Use of Knowledge in Society." *American Economic Review* 35, no. 4 (September 1945): 519–530.

Hayes, Denis. *Rays of Hope: The Transition to a Post-Petroleum World.* New York: W. W. Norton, 1977.

Hayes, Denis. *Worldwatch Paper 4: Energy: The Case for Conservation.* Washington, DC: Worldwatch Institute, January 1976.

Heilbroner, Robert. *An Inquiry into the Human Prospect.* New York: W. W. Norton, 1974.

Henderson, Hazel. *Creating Alternative Futures: The End of Economics.* New York: Berkley Publishing, 1978.

Herbst, Anthony F. *Commodity Futures: Markets, Methods of Analysis, and Management of Risk.* New York: John Wiley and Sons, 1986.

Hines, Lawrence G. *Environmental Issues: Population, Pollution, and Economics.* New York: W. W. Norton, 1973.

Hirschfeld, David J. "A Fundamental Overview of the Energy Futures Markets." *Journal of Futures Markets* (Spring 1983).

Holdren, John, and Philip Herrera. *Energy: A Crisis in Power.* San Francisco: Sierra Club, 1971.

Hubbert, M. King. "Energy from Fossil Fuels." *Science* 109, no. 2823 (February 4, 1949): 104–105.

Hubbert, M. King. *Energy Resources: A Report to the Committee on Natural Resources of the*

National Academy of Sciences–National Research Council. Washington, DC: National Academy of Sciences–National Research Council, 1962.

Hubbert, M. King. *Nuclear Energy and the Fossil Fuels.* Houston, TX: Shell Development Company, June 1956.

Hubbert, M. King. "The World's Evolving Energy System." *American Journal of Physics* 49, no. 1007 (1981): 1026–1027.

Illich, Ivan. *Energy and Equity.* London: Calder and Boyars, 1974.

Illich, Ivan. "Energy and Social Disruption." *Ecologist* 4, no. 2 (February 1974), 49–52.

Illich, Ivan. *Tools for Conviviality.* New York: Harper and Row, 1973.

Jevons, William S. *The Coal Question: An Inquiry Concerning the Progress of the Nation, and the Probable Exhaustion of our Coal-Mines.* London: Macmillan, 1865.

Jickling, Mark. *Futures Markets and the Price of Oil.* Congressional Research Service Report: April 3, 1991.

Joint Committee on Atomic Energy. *Understanding the National Energy Dilemma.* Washington, DC: Government Printing Office, 1973.

Kahn, Herman. *On Thermonuclear War.* Princeton, NJ: Princeton University Press, 1960.

Kahn, Herman, William Morle Brown, and Leon Martel. *The Next 200 Years: A Scenario for America and the World.* New York: Morrow, 1976.

Kahn, Herman, and B. Bruce-Briggs. *Things to Come: Thinking about the Seventies and Eighties.* New York: Macmillan, 1972.

Keynes, John Maynard. *The General Theory of Employment Interest and Money.* London: Macmillan, 1936.

Laird, Melvin R. *Energy: A Crisis in Public Policy.* Washington, DC: American Enterprise Institute, 1977.

Landsberg, Hans, Kenneth J. Arrow, Francis M. Bator, Kenneth W. Dam, Robert W. Fri, Edward R. Fried, Richard L. Garwin, et al. *Energy: The Next Twenty Years: A Report.* Cambridge, MA: Ballinger, 1979.

Landsberg, Hans H., Leonard L. Fischman, and Joseph L. Fisher. *Resources in America's Future: Patterns of Requirements and Availabilities, 1960–2000.* Baltimore: Johns Hopkins University Press, 1963.

Lasch, Christopher. *The Culture of Narcissism: American Life in an Age of Diminishing Expectations.* New York: W. W. Norton, 1979.

Legislative History of the Grain Futures Act of 1922: P.L. 67–311. Washington, DC: Kirkland and Ellis, 1922. US Federal Legislative History Library, www.heinonline.org.

Leopold, Aldo. "The Ecological Conscience." In *The River of the Mother of God and Other Essays by Aldo Leopold*, edited by Susan L. Flader and J. Baird Coolicott. Madison: University of Wisconsin Press, 1991.

Leopold, Aldo. *A Sand County Almanac, and Sketches Here and There.* New York: Oxford University Press, 1949.

Lilienfeld, Robert. *The Rise of Systems Theory: An Ideological Analysis.* New York: John Wiley and Sons, 1978.

Lincoln, G. A. "Energy Conservation." *Science* 180, no. 4082 (Apr. 13, 1973): 161.

Lindeman, Raymond L. "The Trophic-Dynamic Aspect of Ecology." *Ecology* 23, no. 4 (October 1942): 399–417.

Livingston, Robert S., and Truman D. Anderson. *A Desirable Energy Future: A National Perspective*. Philadelphia: Franklin Institute Press, 1982.

Lovins, Amory. *Soft Energy Paths: Toward a Durable Peace*. New York: Harper Colophon Books, 1979.

MacAvoy, Paul W. *Energy Policy: An Economic Analysis*. New York: W. W. Norton, 1983.

MacAvoy, Paul W., and Robert S. Pindyck. *Price Controls and Natural Gas Shortage*. Washington, DC: American Enterprise Institute, 1975.

Malthus, Thomas Robert. *An Essay on the Principle of Population*. Selected and Introduced by Donald Winch. New York: Cambridge University Press, 1992.

Mandel, Leon. *Driven: The American Four-Wheeled Love Affair*. New York: Stein and Day, 1977.

Marsh, George Perkins. *Man and Nature: Or, Physical Geography as Modified by Human Action*. Cambridge, MA: Belknap Press, 2009.

Matthiessen, Peter. *Wildlife in America*. New York: Viking Press, 1959.

McClory, Paul. *How to Use Natural Energy*. National Energy Center, 1978.

Meadows, Donella H., Dennis L. Meadows, Jørgen Randers, and William W. Behrens III. *The Limits to Growth: A Report for the Club of Rome's Project on the Predicament of Mankind*. New York: Universe Books, 1972.

Melamed, Leo, and Bob Tamarkin. *Escape to the Futures*. New York: John Wiley and Sons, 1996.

Mesarovic, Mihajlo, and Eduard Pestel. *Mankind at the Turning Point: The Second Report of the Club of Rome*. New York: E. P. Dutton / Reader's Digest Press, 1974.

Mitchell, Edward J., ed. *Perspectives on U.S. Energy Policy: A Critique of Regulation*. New York: Praeger Publishers, 1976.

Mitchell, Edward J., ed. *Seminar on Energy Policy: The Carter Proposals*. Washington, DC: American Enterprise Institute, 1979.

Mitchell, Edward J. *US Energy Policy: A Primer*. Washington, DC: American Enterprise Institute, 1975.

National Petroleum Council. *U.S. Energy Outlook: An Initial Appraisal, 1971–1985 Volume 1*. Washington, DC: National Petroleum Council, 1971.

National Petroleum Council. *U.S. Energy Outlook*. Washington, DC: National Petroleum Council, 1972.

National Research Council Committee on Nuclear and Alternative Energy Systems. *Energy in Transition 1985–2010*. Washington, DC: National Academy of Sciences, 1979.

New York Mercantile Exchange. *NYMEX Energy Hedging Manual*. New York: New York Mercantile Exchange, 1986.

Nixon, Richard. "Address to the Nation about Policies to Deal with Energy Shortages, November 7, 1973," 916–922. *Public Papers of the Presidents of the United States: Richard Nixon 1973*. Washington, DC: Government Printing Office, 1975.

Nixon, Richard. "Remarks at the Seafarers International Union Biennial Convention," November 26, 1973, 976–981. *Public Papers of the Presidents of the United States: Richard Nixon 1973*. Washington, DC: Government Printing Office, 1975.

Nixon, Richard. "Special Message to the Congress Proposing Energy Emergency

Legislation, November 8, 1973," 922–926. *Public Papers of the Presidents of the United States: Richard Nixon 1973.* Washington, DC: Government Printing Office, 1975.

Odum, Eugene P. "Energy Flow in Ecosystems: A Historical Review." *American Zoologist* 8 (1968): 11–18.

Odum, Howard T. "The Ecosystem, Energy, and Human Values." *Zygon* 12 (June 1977): 109–133.

Odum, Howard T. "Energy, Ecology, and Economics." In "Energy and Society," special issue, *Ambio* 2, no. 6 (1973): 220–227.

Odum, Howard T. *Environment, Power, and Society.* New York: Wiley-Interscience, 1971.

Odum, Howard T. *System Ecology: An Introduction.* New York: Wiley, 1983.

Odum, Howard T., and Elisabeth C. Odum. *Energy Basis for Man and Nature.* New York: McGraw-Hill, 1976.

Ophuls, William. *Ecology and the Politics of Scarcity: Prologue to a Political Theory of the Steady State.* San Francisco: W. H. Freeman, 1977.

Ordway, Samuel H., Jr. *Resources and the American Dream, Including a Theory of the Limit to Growth.* New York: Ronald Press Company, 1953.

Osborn, Fairfield, Jr. *Our Plundered Planet.* Boston: Little Brown, 1948.

Peck, Anne E., ed. *Futures Markets: Their Economic Role.* Washington, DC: American Enterprise Institute for Policy Studies, 1985.

Perlman, Robert, and Roland L. Warren. *Families in the Energy Crisis: Impacts and Implications for Theory and Policy.* Cambridge, MA: Ballinger, 1977.

Pinchot, Gifford. *The Fight for Conservation.* New York: Doubleday, Page, 1910.

Pinchot, Gifford. "Prosperity." In *American Earth: Environmental Writing since Thoreau,* edited by Bill McKibben. New York: Library of America, 2008.

Prast, William G., and Howard L. Lax. *Oil-Futures Markets: An Introduction.* Lexington, MA: Lexington Books, 1983.

President's Materials Policy Commission. *Resources for Freedom: Summary of Volume 1.* Washington, DC: Government Printing Office, June 1952. https://babel.hathitrust .org/cgi/pt?id=uc1.b3897842&view=1up&seq=5.

President's Materials Policy Commission. *Resources for Freedom: Volume III: The Outlook for Energy Sources.* Washington, DC: Government Printing Office, 1952.

Purcell, Arthur H. *The Waste Watchers: A Citizen's Handbook for Conserving Energy and Resources.* Garden City, NY: Anchor Books, 1980.

Razavi, Hossein, and Fereidun Fesharaki. *Fundamentals of Petroleum Trading.* New York: Praeger, 1991.

Resources for the Future. *U.S. Energy Policies: An Agenda for Research.* Washington, DC: United States Office of Science and Technology, 1968.

Rickover, H. G. "Address before Minnesota State Medical Association: Energy Resources and Our Future, May 14, 1957," *Energy Bulletin,* December 2, 2006. http://large .stanford.edu/courses/2011/ph240/klein1/docs/rickover.pdf.

Robertson, James. "Breakdown or Breakthrough: Modern Society at the Turning Point." *Ecologist* 7, no. 9 (November 1977): 340–348.

Rocks, Lawrence, and Richard P. Runyon. *The Energy Crisis.* New York: Crown, 1972.

Rodgers, William. *Brown-Out: The Power Crisis in America.* New York: Stein and Day, 1972.

Rose, Frank S., ed. *Commodity Trading Manual*. Chicago: Board of Trade of the City of Chicago, 1998.

Roszak, Theodore. *The Making of a Counter Culture: Reflections on the Technocratic Society and Its Youthful Opposition*. Garden City, NY: Doubleday, 1969.

Rufus, Miles, Jr. *Awakening from the American Dream*. New York: Universe Books, 1976.

Safer, Arnold E. *International Oil Policy*. Lexington, MA: Lexington Books, 1979.

Scarlott, Charles A. *Energy Sources: Wealth of the World*. New York: McGraw-Hill, 1952.

Schmalz, Anton B., ed. *Energy: Today's Choices, Tomorrow's Opportunities: Essential Dimensions in Thinking for Energy Policy*. Washington, DC: World Future Society, 1974.

Schumacher, E. F. *Small Is Beautiful: Economics As If People Mattered*. New York: Harper and Row, 1975.

Schurr, Sam H., and Bruce C. Netschert, with Vera F. Eliasberg, Joseph Lerner, and Hans H. Landsberg. *Energy in the American Economy, 1850–1975: An Economic Study of Its History and Prospects*. Baltimore: Johns Hopkins University Press, 1960.

Schwager, Jack D. *A Complete Guide to the Futures Markets: Fundamental Analysis, Technical Spread, Trading, Spreads, and Options*. New York: John Wiley and Sons, 1984.

Seevers, Gary L. "Government Regulation and the Futures Markets." *Western Journal of Agricultural Economics* 1, no.1 (June 1977): 21–27.

Simon, Julian L. *The Ultimate Resource*. Princeton, NJ: Princeton University Press, 1981.

Simon, Julian L. *The Ultimate Resource II*. Princeton, NJ: Princeton University Press, 1996.

Simon, Julian, and Herman Kahn, eds. *The Resourceful Earth: A Response to Global 2000*. New York: Blackwell, 1984.

Simon, William E. *A Time for Action*. New York: Reader's Digest Press, 1980.

Snyder, Gary. *Back Country*. New York: New Directions, 1968.

Snyder, Gary. *The Real Work: Interviews and Talks 1964–1979*. New York: New Directions, 1980.

Snyder, Gary. *Turtle Island*. New York: New Directions, 1974.

Soddy, Frederick. *Wealth, Virtual Wealth, and Debt*. 2nd ed. New York: Dutton, 1933.

Statistical Office of the United Nations. *World Energy Supplies*. New York: Statistical Office of the United Nations, 1952–1979.

Stavrianos, L. S. *The Promise of the Coming Dark Age*. San Francisco: W. H. Freeman, 1976.

Stern, Jane. *Trucker: A Portrait of the Last American Cowboy*. New York: McGraw-Hill, 1975.

Stiglitz, Joseph E. "Growth with Exhaustible Natural Resources: The Competitive Economy." *Review of Economic Studies* 41: Symposium on Economics of Exhaustible Resources (1974): 139–152.

Stockes, Bruce. *Worldwatch Paper 17: Local Responses to Global Problems: A Key to Meeting Basic Human Needs*. Washington, DC: Worldwatch Institute, 1978.

Streit, Manfred E., ed. *Futures Markets: Modelling, Managing and Monitoring Futures Trading*. Oxford: Basil Blackwell, 1983.

Talbot, David, and Richard E. Morgan. *Power and Light: Political Strategies for the Solar Transition*. New York: Pilgrim Press, 1981.

Tamarkin, Bob. *The New Gatsbys: Fortunes and Misfortunes of Commodity Traders*. New York: William Morrow, 1985.

Telleen, Maurice. *The Draft Horse Primer: A Guide to the Care and Use of Work Horses and Mules*. Emmaus, PA: Rodale, 1977.

Teller, Edward. *Energy from Heaven and Earth*. San Francisco: W. H. Freeman and Company, 1979.

Teweles, Richard J., Charles V. Harlow, and Herbert L. Stone, eds. *The Commodity Futures Trading Guide: The Science and Art of Sound Commodity Trading*. New York: McGraw-Hill, 1969.

Teweles, Richard J., and Frank J. Jones. *The Futures Game: Who Wins? Who Loses? Why?* New York: McGraw-Hill, 1987.

Todd, Nancy Jack, and John Todd. *Bioshelters, Ocean Arks, City Farming: Ecology as the Basis of Design*. San Francisco: Sierra Club Books, 1984.

Toffler, Alvin. *Future Shock*. New York: Random House, 1970.

Treat, John Elting, ed. *Energy Futures: Trading Opportunities for the 1980s*. Tulsa, OK: PennWell, 1984.

Treat, John Elting, ed. *Energy Futures: Trading Opportunities for the 1990s*. Tulsa, OK: PennWell, 1990.

US Bureau of the Census. *The Statistical History of the United States from Colonial Times to the Present*. New York: Basic Books, 1976.

US Congress. Committee on Agriculture. *A Study of the Effects on the Economy of Trading in Futures and Options*. 98th Cong., 2nd Sess. [Committee Print], (January 1985).

US Congress. Committee on Interior and Insular Affairs. *Fuel and Energy Resources, 1972 Part II*. 92nd Cong., 2nd Sess. April 14, 17, 18, 19, 1972.

US Congress. Committee on Interstate and Foreign Commerce. *Energy Conservation and Oil Policy, Part I: Hearings before the Subcommittee on Energy and Power*. 94th Cong., 1st Sess. March 10–21, May 7, 1975.

US Congress. Committee on Science and Astronautics. *Research Development and the Energy Crisis: Hearing before the Subcommittee on Energy*. 93rd Cong, 1st Sess. November 20, 1973.

US Congress. Joint Economic Committee. *Energy in the Eighties: Can We Avoid Scarcity and Inflation? Hearings before the Subcommittee on Energy*. 95th Cong., 2nd Sess. March 8, 9, 21, 1978. Washington, DC: Government Printing Office, 1978.

US Congress, Senate. Committee on Governmental Affairs. *Oil Prices and Supplies in the Wake of the Persian Gulf Crisis: Hearings before the Committee on Governmental Affairs*. 101st Cong., 2nd sess. October 25, 31, November 1, 1990.

US Congress, Senate. Committee on Homeland Security and Governmental Affairs. *The Role of Market Speculation in Rising Oil and Gas Prices: A Need to Put the Cop Back on the Beat: Staff Report by the Permanent Subcommittee on Investigations*. 109th Cong., 2nd Sess. June 27, 2006.

US Congress, Senate. Committee on Homeland Security and Governmental Affairs and Committee on Energy and Natural Resources. *Speculation in the Crude Oil Market: Joint Hearing before the Permanent Subcommittee on Investigations and the Subcommittee on Energy*. 110th Cong., 1st Sess. December 11, 2007.

US Congress. Testimony of Michael W. Masters before the Committee on Homeland Security and Governmental Affairs, United States Senate. May 20, 2008. https://www.hsgac.senate.gov/imo/media/doc/052008Masters.pdf?attempt=2.

US Department of Energy. *Securing America's Energy Future: The National Energy Policy Plan*. Washington, DC: U.S. Department of Energy, 1981.

US Department of State. *Energy Resources of the World*. Washington, DC: Government Printing Office, 1949.

US Federal Power Commission. *The 1970 National Power Survey*. Washington, DC: Federal Power Commission, 1970.

Vacca, Roberto. *The Coming Dark Age*. Garden City, NY: Anchor Press–Doubleday, 1974.

Verleger, Philip A., Jr. "The Potential Impacts of Trading in Oil Futures on the World Oil Market." In *Energy Futures: Trading Opportunities for the 1980s*, edited by John Elting Treat, 114–123. Tulsa, OK: PennWell, 1984.

Vitiello, Jane Kagan. *Trading through Time: The History of the New York Mercantile Exchange, 1872–1997*. New York: New York Mercantile Exchange, 1997.

Vogt, William. *Road to Survival*. New York: W. Sloane Associates, 1948.

Weidenbaum, Murray L., Reno Harnish, and James McGowen. *Government Credit Subsidies for Energy Development*. Washington, DC: American Enterprise Institute, 1976.

White, Leslie A. "Energy and the Evolution of Culture." *American Anthropologist* 45, no. 3 (July–September 1943): 335–356.

White, Leslie A. *The Science of Culture: A Study of Man and Civilization*. New York: Grove Press, 1949.

Whitman, Walt. *Leaves of Grass: The Complete 1855 and 1891–92 Editions*. New York: Library of American Paperback Classics, 2011.

World Bank. *The Energy Transition in Developing Countries*. Washington, DC: International Bank for Development and Reconstruction, 1983.

Yates, Brock. *Cannonball! World's Greatest Outlaw Road Race*. Saint Paul, MN: Motorbooks, 2003.

Yergin, Daniel, and Martin Hillenbrand. *Global Insecurity: A Strategy for Energy and Economic Renewal*. Boston: Houghton Mifflin, 1982.

Yergin, Daniel, and Robert Stobaugh, eds. *Energy Future: Report of the Energy Project at the Harvard Business School*. New York: Random House, 1979.

Filmography

American Graffiti. Directed by George Lucas. Universal Pictures and Lucasfilm, 1973.

Badlands. Directed by Terrence Malick. Warner Bros., 1973.

Battletruck. Directed by Harley Cokeliss. Battletruck Films, 1982.

The Betsy. Directed by Daniel Petrie. Harold Robbins International Company, 1978.

BJ and the Bear. Created by Glen A. Larson and Christopher Crowe. Universal Television, 1978–81.

The Blues Brothers. Directed by John Landis. Universal Pictures, 1980.

Breaker Breaker! Directed by Don Hulette. Paragon Films, 1977.

Bullitt. Directed by Peter Yates. Solar Productions, 1968.

The California Kid. Directed by Richard T. Heffron. Universal Television, 1974.

Cannonball! Directed by Paul Bartel. New World Pictures, 1976.

The Cannonball Run. Directed by Hal Needham. Golden Harvest Company, 1981.

The Cannonball Run II. Directed by Hal Needham. Golden Harvest Company and Warner Bros., 1984.

The Car. Directed by Elliot Silverstein. Universal Pictures, 1977.

Christine. Directed by John Carpenter. Columbia, 1983.

Convoy. Directed by Sam Peckinpah. EMI Films and United Artists, 1978.

Corvette Summer. Directed by Matthew Robbins. Metro-Goldwyn-Mayer (MGM), 1978.

Crazy Mama. Directed by Jonathan Demme. New World Pictures, 1975.

Damnation Alley. Directed by Jack Smight. Twentieth Century–Fox, 1977.

Deadhead Miles. Directed by Vernon Zimmerman. Paramount Pictures, 1973.

Deadline Auto Theft. Directed by H. B. Halicki. H. B. Halicki Mercantile Co., 1983.

Death Race 2000. Directed by Paul Bartel. New World Pictures, 1975.

Deathsport. Directed by Allan Arkush and Nicholas Niciphor. New World Pictures, 1978.

Dirty Mary Crazy Larry. Directed by John Hough. Twentieth Century–Fox and Academy Pictures Corp., 1974.

Don't Be Fuelish Public Service Announcements. Advertising Council, 1974ff.

Don't Blow It America Public Service Announcements. Alliance to Save Energy, 1979–80.

Double Nickels. Directed by Jack Vacek. Smokey Productions Co., 1977.

The Driver. Directed by Walter Hill. EMI Films and Twentieth Century–Fox, 1978.

Duel. Directed by Steven Spielberg. Universal Television, 1971.

Easy Rider. Directed by Dennis Hopper. Pando Company, 1969.

Eat My Dust. Directed by Charles B. Griffith. New World Pictures, 1976.

The Fast and the Furious. Directed by John Ireland and Edward Sampson. Palo Alto Productions, 1955.

Fast Charlie . . . the Moonbeam Rider. Directed by Steve Carver. Universal Pictures, 1979.

Fast Company. Directed by David Cronenberg. Quadrant Films, 1979.

Five Easy Pieces. Directed by Bob Rafelson. BBS Productions, 1970.

The Formula. Directed by John G. Avildsen. MGM, 1980.

The French Connection. Directed by William Friedkin. D'Antoni Productions, 1971.

Gator. Directed by Burt Reynolds. Levy-Gardner-Laven, 1976.

The Gauntlet. Directed by Clint Eastwood. Warner Bros., 1977.

Gone in 60 Seconds. Directed by H. B. Halicki. H. B. Halicki Mercantile Co., 1974.

The Graduate. Directed by Mike Nichols. Lawrence Turman, 1967.

Grand Prix. Directed by John Frankenheimer. MGM, 1966.

Grand Theft Auto. Directed by Ron Howard. New World Pictures, 1977.

Greased Lightning. Directed by Michael Schultz. Third World Cinema and Warner Bros., 1977.

The Great Smokey Roadblock. Directed by John Leone. Cinema Arts Assoc., 1977.

The Gumball Rally. Directed by Charles Bail. First Artists and Warner Bros., 1976.

The Hearse. Directed by George Bowers. Marimark Productions, 1980.

High Ballin'. Directed by Peter Carter. AIP, 1978.

Hooper. Directed by Hal Needham. Warner Bros., 1978.

Hot Rod. Directed by George Armitage. ABC Circle Films, 1979.

The Italian Job. Directed by Peter Collinson. Oakhurst Productions, 1969.

Jackson County Jail. Directed by Michael Miller. New World Pictures, 1976.

The Junkman. Directed by H. B. Halicki. H. B. Halicki Mercantile, 1982.

King Kong. Directed by John Guillermin. Dino De Laurentiis Company, 1976.

Logan's Run. Directed by Michael Anderson. MGM, 1976.

Macon County Line. Directed by Richard Compton. Max Baer Productions, 1974.

Mad Max. Directed by George Miller. Kennedy Miller Productions, 1979.

Moonrunners. Directed by Gy Waldron. United Artists, 1975.

Mother, Jugs, and Speed. Directed by Peter Yates. Twentieth Century–Fox, 1976.

Movin' On. Created by Philip D'Antoni and Barry J. Weitz. D'Antoni/Weitz Productions, 1974–76.

Network. Directed by Sidney Lumet. MGM, 1976.

The New Alchemists. Directed by Dorothy Todd Hénaut. National Film Board of Canada, 1974.

Rollerball. Directed by Norman Jewison. Algonquin and United Artists, 1975.

The Seven Ups. Directed by Philip D'Antoni. Twentieth Century–Fox, 1973.

Silent Running. Directed by Douglas Trumbull. Universal Pictures, 1972.

Smokey and the Bandit. Directed by Hal Needham. Universal Pictures, 1977.

Smokey and the Bandit II. Directed by Hal Needham. Universal Pictures, 1980.

Sorcerer. Directed by William Friedkin. Paramount Pictures, 1977.

Soylent Green. Directed by Richard Fleischer. MGM, 1973.

Steel Cowboy. Directed by Harvey S. Laidman. EMI Films, 1978.

Stingray. Directed by Richard Taylor. Stingray Productions, 1978.

Stroker Ace. Directed by Hal Needham. Universal Pictures, 1983.

The Sugarland Express. Directed by Steven Spielberg. Universal Pictures, 1974.

Texas Lightning. Directed by Gary Graver. Productions Two, 1981.

Three Days of the Condor. Directed by Sydney Pollack. Wildwood Enterprises and Dino De Laurentiis Company, 1975.

Thunder and Lightning. Directed by Corey Allen. Twentieth Century–Fox, 1977.

Thunderbolt and Lightfoot. Directed by Michael Cimino. Malpaso Company, 1974.

Truck Stop Women. Directed by Mark L. Lester. L-T Films, 1974.

Two-Lane Blacktop. Directed by Monte Hellman. Universal Pictures, 1971.

Vanishing Point. Directed by Richard C. Sarafian. Cupid Productions, 1971.

White Lightning. Directed by Joseph Sargent. Levy-Gardner-Laven, 1973.

White Line Fever. Directed by Jonathan Kaplan. Columbia, 1975.

W. W. Dixie and the Dancekings. Directed by John G. Avildsen. Twentieth Century–Fox, 1975.

ZPG. Directed by Michael Campus. Sagittarius Productions, 1972.

Secondary Sources
Books, Articles, and Media

Adams, Vincanne, Michelle Murphy, and Adele Clarke. "Anticipation." *Subjectivity* 28 (2009): 246–265.

Adelman, Morris. *The Genie Out of the Bottle: World Oil Since 1970*. Cambridge, MA: MIT Press, 1996.

Aguilera, Roberto F., and Marian Radetzki. *The Price of Oil*. New York: Cambridge University Press, 2016.

Alizadeh, Parvin. "The Political Economy of Petro Populism and Reform, 1997–2011." In *Iran and the Global Economy: Petro Populism, Islam, and Economic Sanctions*, edited by Parvin Alizadeh and Hassan Hakimian, 76–101. Abingdon, Oxon: Routledge, 2014.

Anderson, David M. "Levittown Is Burning! The 1979 Levittown, Pennsylvania, Gas Line Riot and the Decline of the Blue-Collar American Dream." *Labor: Studies in Working-Class History of the Americas* 2, no. 3 (2005): 47–65.

Anderson, Ray C., ed. *Berkshire Encyclopedia of Sustainability*. Great Barrington, MA: Berkshire Publishing Group, 2012.

Avila, Eric. *Popular Culture in the Age of White Flight: Fear and Fantasy in Suburban Los Angeles*. Berkeley: University of California Press, 2004.

Bakke, Gretchen. *The Grid: The Fraying Wires between Americans and Our Energy Future*. New York: Bloomsbury, 2016.

Barber, Daniel A. *A House in the Sun: Modern Architecture and Solar Energy in the Cold War*. New York: Oxford University Press, 2016.

Barrctt, Ross, and Daniel Worden, eds. *Oil Culture*. Minneapolis: University of Minnesota Press, 2014.

Beckert, Jens. *Imagined Futures: Fictional Expectations and Capitalism Dynamics*. Cambridge, MA: Harvard University Press, 2016.

Binkley, Sam. *Getting Loose: Lifestyle Consumption in the 1970s*. Durham, NC: Duke University Press, 2007.

Black, Brian. *Crude Reality: Petroleum in World History*. New York: Rowman and Littlefield, 2012.

Black, Brian. "Energizing Environmental History." In *A Field on Fire: The Future of Environmental History*, edited by Mark D. Hersey and Ted Steinberg. Tuscaloosa: University of Alabama Press, 2019.

Bliss, Michael. *Justified Lives: Morality and Narrative in the Films of Sam Peckinpah*. Carbondale: Southern Illinois University Press, 1993.

Borstelmann, Thomas. *The 1970s: A New Global History from Civil Rights to Economic Inequality*. Princeton, NJ: Princeton University Press, 2012.

Boyer, Dominic. "Energopower: An Introduction." *Anthropological Quarterly: Energopower and Biopower in Transition* 87, no. 2 (2014): 309–334.

Bramwell, Anna. *Ecology in the Twentieth Century: A History*. New Haven: Yale University Press, 1989.

Brown, Adrienne, and Valerie Smith. *Race and Real Estate*. New York: Oxford University Press, 2015.

Brown, Wendy. "American Nightmare: Neo-liberalism, Neo-conservatism, and De-Democratization." *Political Theory* 34, no. 6 (December 2006): 690–714.

Brown, Wendy. *Undoing the Demos: Neoliberalism's Stealth Revolution*. New York: Zone Books, 2015.

Buell, Frederick. *From Apocalypse to Way of Life: Environmental Crisis in the American Century.* New York: Routledge, 2003.

Carlisle, Juliet E., Jessica T. Feezell, Kristy E. H. Michaud, and Eric R. A. N. Smith. *The Politics of Energy Crises.* New York: Oxford University Press, 2017.

Carter, Dan T. *From George Wallace to Newt Gingrich: Race in the Conservative Counterrevolution, 1963–1994.* Baton Rouge: Louisiana State University Press, 1996.

Clark, John G. *Energy and the Federal Government: Fossil Fuel Policies, 1900–1946.* Urbana: University of Illinois Press, 1987.

Clarke, Duncan. *The Battle for Barrels: Peak Oil Myths and World Oil Futures.* London: Profile Books, 2007.

Clayton, Blake C. *Market Madness: A Century of Oil Panics, Crises, and Crashes.* New York: Oxford University Press, 2015.

Cohen, Lizabeth. *A Consumers' Republic: The Politics of Mass Consumption in Postwar America.* New York: Vintage Books, 2003.

Coll, Steve. *Private Empire: ExxonMobil and American Power.* New York: Penguin, 2012.

Connelly, Matthew. *Fatal Misconception: The Struggle to Control World Population.* Cambridge, MA: Belknap Press, 2008.

Cook, David. *Lost Illusions: American Cinema in the Shadow of Watergate and Vietnam.* Berkeley: University of California Press, 2000.

Cook, Robert Edward. "Raymond Lindeman and the Trophic-Dynamic Concept in Ecology." *Science* 198, no. 4312 (October 7, 1977): 22–26.

Cooper, Gail. *Air-Conditioning America: Engineers and the Controlled Environment, 1900–1960.* Baltimore: Johns Hopkins University Press, 2002.

Cooper, Melinda. *Family Values: Between Neoliberalism and the New Social Conservatism.* New York: Zone Books, 2017.

Cooper, Melinda. *Life as Surplus: Biotechnology and Capitalism in the Neoliberal Era.* Seattle: University of Washington Press, 2008.

Corkin, Stanley. *Starring New York: Filming the Grime and Glamour of the Long 1970s.* New York: Oxford University Press, 2011.

Costanza, Robert, John H. Cumberland, and Herman Daly, eds. *Introduction to Ecological Economics.* 2nd ed. New York: CRC Press, 2015.

Cowie, Jefferson. *The Great Exception: The New Deal and the Limits of American Politics.* Princeton, NJ: Princeton University Press, 2016.

Crabb, Cecil V., Jr. "The Energy Crisis, the Middle East, and American Foreign Policy," *World Affairs* 136 (Summer 1973): 48–73.

Crespino, Joseph. *In Search of Another Country: Mississippi and the Conservative Counterrevolution.* Princeton, NJ: Princeton University Press, 2007.

Cronon, William. *Nature's Metropolis: Chicago and the Great West.* New York: W. W. Norton, 1991.

Curtis, Neal. *Idiotism: Capitalism and the Privatisation of Life.* London: Pluto Press, 2013.

Dagget, Cara New. *The Birth of Energy: Fossil Fuels, Thermodynamics, and the Politics of Work.* Durham, NC: Duke University Press, 2019.

Debeir, Jean-Claude, Jean-Paul Delége, and Daniel Hémery. *In the Servitude of Power:*

Energy and Civilisation through the Ages. Translated by John Barzman. London: Zed Books, 1991.

Decker, Mark T. "They Want Unfreedom and One-Dimensional Thought? I'll Give Them Unfreedom and One-Dimensional Thought: George Lucas, *THX-1138*, and the Persistence of the Marcusian Social Critique in *American Graffiti* and the *Star Wars* Films." *Extrapolation* 50, no. 3 (2009): 417–441.

Deffeyes, Kenneth S. *Hubbert's Peak: The Impending World Oil Shortage*. Princeton, NJ: Princeton University Press, 2001.

Deffeyes, Kenneth S. *When Oil Peaked*. New York: Hill and Wang, 2010.

De Rycker, Antoon, and Zuraidah Mohd Don. Introduction to *Discourse and Crisis: Critical Perspectives*, 6–7. Edited by Antoon De Rycker and Zuraidah Mohd Don. Philadelphia: John Benjamins Publishing, 2013.

de Vogli, Robert. *Progress or Collapse: The Crises of Market Greed*. New York: Routledge, 2013.

Dochuk, Darren. *From Bible Belt to Sunbelt: Plain-Folk Religion, Grassroots Politics, and the Rise of Evangelical Conservatism*. New York: W. W. Norton, 2012.

Dumenil, Gerard, and Dominique Levy. *The Crisis of Neoliberalism*. Cambridge, MA: Harvard University Press, 2011.

Dunaway, Finis. "Gas Masks, Pogo, and the Ecological Indian: Earth Day and the Visual Politics of American Environmentalism." *American Quarterly* 60, no. 1 (March 2008): 67–99.

Dunaway, Finis. *Seeing Green: The Use and Abuse of American Environmental Images*. Chicago: University of Chicago Press, 2015.

Egan, Michael. *Barry Commoner and the Science of Survival*. Cambridge, MA: MIT Press, 2007.

Elsaesser, Thomas, Alexander Horwath, and Noel King, eds. *The Last Great American Picture Show: New Hollywood Cinema in the 1970s*. Amsterdam: Amsterdam University Press, 2004.

Fabian, Ann. *Card Sharps, Dream Books, and Bucket Shops: Gambling in 19th Century America*. Ithaca, NY: Cornell University Press, 1990.

Farish, Matthew. *The Contours of America's Cold War*. Minneapolis: University of Minnesota Press, 2010.

Fiege, Mark. *The Republic of Nature: An Environmental History of the United States*. Seattle: University of Washington Press, 2013.

Flint, Anthony. *Wrestling with Moses: How Jane Jacobs Took on New York's Master Builder and Transformed an American City*. New York: Random House, 2011.

Flippen, J. Brooks. *Jimmy Carter, the Politics of the Family, and the Rise of the Religious Right*. Athens, GA: University of Georgia Press, 2011.

Fones-Wolf, Elizabeth A., and Ken Fones-Wolf, *Struggle for the Soul of the Postwar South: White Evangelical Protestants and Operation Dixie*. Chicago: University of Chicago Press, 2015.

Foucault, Michel. *The Birth of Biopolitics: Lectures at the College de France, 1978–1979*. Translated by Graham Burchell. New York: Picador, 2004.

Freund, David M. P. *Colored Property: State Policy and White Racial Politics in Suburban America*. Chicago: University of Chicago Press, 2007.

Ganapathy, R. S. "Transportation and Equity: A Review." *Economic and Political Weekly* 10, no. 30 (July 26, 1975): 1117–1118.

Garber, Peter M. "The Collapse of the Bretton Woods Fixed Exchange Rate." In *A Retrospective on the Bretton Woods System: Lessons for International Monetary Reform*, edited by Michael D. Bordo and Barry Eichengreen, 461–485. Chicago: University of Chicago Press, 1993.

Gerstle, Gary. "The Reach and Limits of the Liberal Consensus." In *The Liberal Consensus Reconsidered*, edited by Robert Mason and Iwan Morgan. Gainesville: University Press of Florida, 2017.

Gold, Russell. *The Boom: How Fracking Ignited the American Energy Revolution and Changed the World*. New York: Simon and Schuster, 2014.

Golinksi, Jan. *Making Natural Knowledge: Constructivism and the History of Science*. Chicago: University of Chicago Press, 2005.

Golley, Frank Benjamin. *A History of the Ecosystem Concept: More Than the Sum of Its Parts*. New Haven, CT: Yale University Press, 1993.

Goodman, Leah McGrath. *The Asylum: The Truth about the Renegades Who Stole the World's Oil Market*. New York: William Morrow, 2011.

Gottlieb, Robert. *Forcing the Spring: The Transformation of the American Environmental Movement*. Washington, DC: Island Press, 2005.

Graf, Rüdiger. "Between 'National' and 'Human' Security: Energy Security in the United States and Western Europe in the 1970s." *Historical Social Research* 35, no. 4 (2010): 329–348.

Graf, Rüdiger. "Making Use of the 'Oil Weapon': Western Industrialized Countries and Arab Petropolitics in 1973–74." *Diplomatic History* 36, no. 1 (January 2012): 185–208.

Hagen, Joel B. *An Entangled Bank: The Origins of Ecosystem Ecology*. New Brunswick, NJ: Rutgers University Press, 1992.

Hagen, Joel B. "Eugene and Howard Odum." *Oxford Bibliographies in Ecology*. https://www.oxfordbibliographies.com/view/document/obo-9780199830060/obo-9780199830060-0190.xml#obo-9780199830060-0190-div1-0005.

Hagen, Joel B. "Teaching Ecology During the Environmental Age, 1965–1980." *Environmental History* 13, no. 4 (October 2008): 704–723.

Haiven, Max. *Cultures of Financialization: Fictitious Capital in Popular Culture and Everyday Life*. New York: Palgrave Macmillan, 2014.

Hamblin, Jacob Darwin. *Arming Mother Earth: The Birth of Catastrophic Environmentalism*. New York: Oxford University Press, 2013.

Hamilton, Shane. *Trucking Country: The Road to America's Wal-Mart Economy*. Princeton, NJ: Princeton University Press, 2008.

Hansen, Per H. "From Finance Capitalism to Financialization: A Cultural and Narrative Perspective on 150 Years of Financial History." *Enterprise and Society* 15 (December 2014): 605–642.

Harvey, David. *A Brief History of Neoliberalism*. New York: Oxford University Press, 2007.

Hayden, Dolores. *Building Suburbia: Green Fields and Urban Growth, 1820–2000*. New York: Pantheon Books, 2003.

Heinberg, Richard. *Peak Everything: Waking Up to a Century of Declines*. Gabriola Island, BC: New Society Publishers, 2007.

Heller, Walter. "Kennedy's Supply-Side Economics." *Challenge* 24, no. 2 (May 1981): 14–18.

Helvarg, David. *The War against the Greens: The 'Wise Use' Movement, the New Right, and the Browning of America*. 2nd ed. Boulder: Johnson Books, 2004.

Henriksen, Margot. *Dr. Strangelove's America: Society and Culture in the Atomic Age*. Berkeley: University of California Press, 1997.

Heynen, Nik, James McCarthy, Scott Prudham, and Paul Robbins, eds. *Neoliberal Environments: False Promises and Unnatural Consequences*. New York: Routledge, 2010.

Hieronymus, Thomas A. *Economics of Futures Trading for Commercial and Personal Profit*. 2nd ed. New York: Commodity Research Bureau, 1977.

Hitchcock, Peter. "Oil in an American Imaginary." *New Formations* 69 (March 2010): 81–97.

Ho, Karen. *Liquidated: An Ethnography of Wall Street*. Durham, NC: Duke University Press, 2009.

Hohle, Randolph. *Race and the Origins of American Neoliberalism*. New York: Routledge, 2015.

Horowitz, Daniel. *The Anxieties of Affluence: Critiques of American Consumer Culture, 1939–1979*. Amherst: University of Massachusetts Press, 2005.

Huber, Matthew T. *Lifeblood: Oil, Freedom, and the Forces of Capital*. Minneapolis: University of Minnesota Press, 2013.

Hyman, Louis. *Debtor Nation: A History of America in Red Ink*. Princeton, NJ: Princeton University Press, 2011.

Jackson, Kenneth T. *Crabgrass Frontier: The Suburbanization of the United States*. New York: Oxford University Press, 1985.

Jacobs, Meg. *Panic at the Pump: The Energy Crisis and the Transformation of American Politics in the 1970s*. New York: Hill and Wang, 2016.

Jameson, Fredric. *The Cultural Turn: Selected Writings on the Postmodern, 1983–1998*. New York: Verso, 1998.

Jameson, Fredric. "Culture and Finance Capital." *Critical Inquiry* 24, no. 1 (Autumn 1997): 246–265.

Janiewski, Dolores. "Review: Towards a New Suburban History." *Social History* 33, no. 1 (February 2008): 60–67.

Johnson, Bob. *Carbon Nation: Fossil Fuels in the Making of American Culture*. Lawrence: University Press of Kansas, 2014.

Jones, Christopher F. *Routes of Power: Energy and Modern America*. Cambridge, MA: Harvard University Press, 2016.

Jones, Daniel Stedman. *Masters of the Universe: Hayek, Friedman, and the Birth of Neoliberal Politics*. Princeton, NJ: Princeton University Press, 2012.

Jundt, Thomas. *Greening the Red, White, and Blue: The Bomb, Big Business, and Consumer Resistance in Postwar America*. New York: Oxford University Press, 2014.

Kalman, Laura. *Right Star Rising: A New Politics, 1974–1980*. New York: W. W. Norton, 2010.

Kern, Stephen. *The Culture of Time and Space, 1880–1918*. Cambridge, MA: Harvard University Press, 1983.

Kingsland, Sharon E. *The Evolution of American Ecology, 1890–2000*. Baltimore: Johns Hopkins University Press, 2005.

Kohn, Vilma L. "The Oil Import Question: Research, Report, Reaction." *Case Western Reserve Journal of International Law* 3, no. 1 (1970): 88–105.

Koselleck, Reinhart. "Crisis." Translated by Michaela W. Richter. *Journal of the History of Ideas* (April 2006): 357–400.

Koselleck, Reinhart. *Futures Past: On the Semantics of Historical Time*. Translated by Keith Tribe. New York: Columbia University Press, 2004.

Koselleck, Reinhart. *The Practice of Conceptual History: Timing, History, Spacing Concepts*. Stanford, CA: Stanford University Press, 2002.

Kruse, Kevin M. *White Flight: Atlanta and the Making of Modern Conservatism*. Princeton, NJ: Princeton University Press, 2005.

Kuhn, Thomas. *The Structure of Scientific Revolutions*. Chicago: University of Chicago Press, 1996.

Labban, Mazen. "Oil in Parallax: Scarcity, Markets, and the Financialization of Accumulation." *Geoforum* 41 (2010): 541–552.

Lacalle, Daniel, and Diego Parrilla. *The Energy World Is Flat: Opportunities from the End of Peak Oil*. New York: Wiley, 2015.

Laderman, David. *Driving Visions: Exploring the Road Movie*. Austin: University of Texas Press, 2002.

Lakoff, Andrew. "From Population to Vital Systems: National Security and the Changing Object of Public Health." In *Biosecurity Interventions: Global Health and Security in Question*, edited by Stephen J. Collier and Andrew Lakoff. New York: Columbia University Press, 2008.

Lassiter, Matthew D. "Political History beyond the Red-Blue Divide." *Journal of American History* 98, no. 3 (December 2011): 760–764.

Lassiter, Matthew D. *Silent Majority: Suburban Politics in the Sunbelt South*. Princeton, NJ: Princeton University Press, 2006.

Lear, Linda J. "Rachel Carson's 'Silent Spring.'" *Environmental History Review* 17, no. 2 (Summer 1993): 23–48.

Leech, Garry. *Crude Interventions: The United States, Oil, and the New World (Dis)Order*. New York: Zed Books, 2006.

Levitt, Kari Polanyi. *From the Great Transformation to the Great Financialization: On Karl Polanyi and Other Essays*. Halifax: Fernwood Publishers, 2013.

Levy, Jonathan. *Freaks of Fortune: The Emerging World of Capitalism and Risk in America*. Cambridge, MA: Harvard University Press, 2012.

Lichtman, Allan J. *White Protestant Nation: The Rise of the American Conservative Movement*. New York: Grove Press, 2008.

Liftset, Robert D. "A New Understanding of the American Energy Crisis of the 1970s." *Historical Social Research/Historische Sozialforschung* 39, no. 4, (2014): 22–42.

Lower, Robert C. "The Regulation of Commodity Options." *Duke Law Journal* (December 1978): 1095–1145.

Mackin, Anne. *Americans and Their Land: The House Built on Abundance.* Ann Arbor: University of Michigan Press, 2006.

Maher, Neil M. *Nature's New Deal: The Civilian Conservation Corps and the Roots of the American Environmental Movement.* New York: Oxford University Press, 2008.

Malm, Andreas. *Fossil Capital: The Rise of Steam and the Roots of Global Warming.* New York: Verso, 2016.

Malm, Andreas. "Who Lit This Fire: Approaching the History of the Fossil Economy." *Critical Historical Studies* 3, no. 2 (Fall 2016): 215–248.

Mancke, Richard B. *Squeaking By: U.S. Energy Policy since the Embargo.* New York: Columbia University Press, 1976.

Markham, Jerry W. *The History of Commodity Futures Derivatives.* New York: Praeger, 1987.

Marling, Karal Ann. *As Seen on TV: The Visual Culture of Everyday Life in the 1950s.* Cambridge, MA: Harvard University Press, 1994.

Marshall, Elaine. "'We're Always Moving': Sam Peckinpah's Making of *Convoy*." In *Sam Peckinpah's West: New Perspectives*, edited by Leonard Engel. Salt Lake City: University of Utah Press, 2003.

Martin, Randy. *An Empire of Indifference: American War and the Financial Logic of Risk.* Durham, NC: Duke University Press, 2007.

Martin, Randy. *Financialization of Daily Life.* Philadelphia: Temple University Press, 2002.

Masco, Joseph. *The Theater of Operations: National Security Affect from the Cold War to the War on Terror.* Durham: Duke University Press, 2014.

Mason, Carol. *Reading Appalachia from Left to Right: Conservatives and the 1974 Kanawha County Textbook Controversy.* Ithaca, NY: Cornell University Press, 2009.

Massumi, Brian. "The Future Birth of the Affective Fact: The Political Ontology of Threat." In *The Affect Theory Reader*, edited by Melissa Gregg and Gregory J. Seigworth. Durham, NC: Duke University Press, 2010.

Mattson, Kevin. *What the Heck Are You Up to Mr. President? Jimmy Carter, America's Malaise, and the Speech That Should Have Changed the Country.* New York: Bloomsbury, 2009.

Matusow, Allen J. *Nixon's Economy: Booms, Busts, Dollars, and Votes.* Lawrence: University Press of Kansas, 1998.

Maugeri, Leonardo. *The Age of Oil: The Mythology, History, and Future of the World's Most Controversial Resource.* New York: Praeger, 2006.

May, Elaine Tyler. *Homeward Bound: American Families in the Cold War Era.* New York: Basic Books, 2008.

McAlister, Melani. *Epic Encounters: Culture, Media, and U.S. Interests in the Middle East since 1945.* Updated ed. Berkeley: University of California Press, 2005.

McGirr, Lisa. *Suburban Warriors: The Origins of the New American Right.* Princeton, NJ: Princeton University Press, 2001.

McGuigan, Jim. *Neoliberal Culture.* New York: Palgrave Macmillan, 2016.

McKillop, Andrew, and Sheila Newman, eds. *The Final Energy Crisis*. Ann Abor, MI: Pluto Press, 2005.

McNally, Robert. *Crude Volatility: The History and the Future of Boom-Bust Oil Prices*. New York: Columbia University Press, 2017.

McNeill, J. R. *Something New Under the Sun: An Environmental History of the Twentieth-Century World*. New York: W. W. Norton, 2000.

McNeill, J. R., and Peter Engelke. *The Great Acceleration: An Environmental History of the Anthropocene since 1945*. Cambridge, MA: Belknap Press, 2016.

McNeill, J. R., Jose Augusto Padua, and Mahesh Rangarajan, eds. *Environmental History as If Nature Existed*. New York: Oxford University Press, 2010.

McQuaig, Linda. *It's the Crude, Dude: War, Big Oil, and the Fight for the Planet*. Toronto: Doubleday Canada, 2004.

Mills, Kate. *The Road Story and the Rebel: Moving through Film, Fiction, and Television*. Carbondale: Southern Illinois University Press, 2006.

Mills, Robin M. *The Myth of the Oil Crisis: The Coming Challenges of Depletion, Geopolitics, and Global Warming*. Westport, CT: Praeger, 2008.

Mirowski, Philip. *Never Let a Serious Crisis Go to Waste: How Neoliberalism Survived the Financial Meltdown*. London: Verso, 2013.

Mirowski, Philip, and Dieter Plehwe, eds. *The Road from Mont Pelerin: The Making of the Neoliberal Thought Collective*. Cambridge, MA: Harvard University Press, 2009.

Mirzoeff, Nicholas. "Visualizing the Anthropocene." *Public Culture* 26, no. 2 (Spring 2014): 213–232.

Mitchell, Timothy. "Carbon Democracy." *Economy and Society* 38, no. 3 (August 2009): 399–432.

Mitchell, Timothy. *Carbon Democracy: Political Power in the Age of Oil*. New York: Verso, 2011.

Mitchell, Timothy. "Fixing the Economy." *Cultural Studies* 12, no. 1 (January 1998): 82–101.

Mitman, Gregg. *The State of Nature: Ecology, Community, and American Social Thought, 1900–1950*. Chicago: University of Chicago Press, 1992.

Moreton, Bethany. *To Serve God and Wal-Mart: The Making of Christian Free Enterprise*. Cambridge, MA: Harvard University Press, 2009.

Morton, Timothy. *Hyperobjects: Philosophy and Ecology after the End of the World*. Saint Paul: University of Minnesota Press, 2013.

Mottram, Eric. "Blood on the Nash Ambassador: Cars in American Films." In *Autopia: Cars and Culture*, edited by Peter Wollen and Joel Kerr. London: Reaktion Books, 2002.

Murphy, Michelle. *Sick Building Syndrome and the Problem of Uncertainty: Environmental Politics, Technoscience, and Women Workers*. Durham, NC: Duke University Press, 2006.

Murphy, Patrick D., ed. *Critical Essays on Gary Snyder*. Boston: C. K. Hall, 1990.

Nash, Roderick. *Wilderness and the American Mind*. 5th ed. New Haven, CT: Yale University Press, 2014.

Nikiforuk, Andrew. *The Energy of Slaves: Oil and the New Servitude*. Toronto: Greystone Books, 2012.

Nikiforuk, Andrew. "The Social Price of Energy." *Alternatives Journal* (November 2012). http://www.alternativesjournal.ca/people-and-profiles/social-price-energy.

Nye, David E. *Consuming Power: A Social History of American Energies.* Cambridge, MA: MIT Press, 1998.

Nystrom, Derek. *Hard Hats, Rednecks, and Macho Men: Class in 1970s American Cinema.* New York: Oxford University Press, 2009.

Omi, Michael, and Howard Winant. *Racial Formation in the United States.* 3rd ed. New York: Routledge, 2015.

O'Sullivan, Mary. "The Expansion of the U.S. Stock Market, 1885–1930: Historical Facts and Theoretical Fictions." *Enterprise and Society* 8, no. 3 (September 2007): 489–542.

Ott, Julia C. *When Wall Street Met Main Street: The Quest for an Investor's Democracy.* Cambridge, MA: Harvard University Press, 2011.

Panitch, Leo, and Sam Gindin. *The Making of Global Capitalism: The Political Economy of American Empire.* New York: Verso, 2013.

Parra, Francisco. *Oil Politics: A Modern History of Petroleum.* New York: I. B. Tauris, 2009.

Partanen, Rauli, Harri Paloheimo, and Keikki Waris. *The World after Cheap Oil.* New York: Routledge, 2015.

Petrocultures Research Group. *After Oil.* Morgantown: West Virginia University Press, 2016.

Phillips-Fein, Kim. "Conservatism: A State of the Field." *Journal of American History* 93, no. 3 (December 2011): 723–743.

Phillips-Fein, Kim. *Invisible Hands: The Businessmen's Crusade against the New Deal.* New York: W. W. Norton, 2009.

Pindyck, Robert S. "The Dynamics of Commodity Spot and Futures Markets: A Primer." *Energy Journal* 22, no. 3 (2001), 1–29.

Price-Smith, Andrew T. *Oil, Illiberalism, and War.* Cambridge, MA: MIT Press, 2015.

Princen, Thomas, Jack P. Manno, and Pamela L. Martin, eds. *Ending the Fossil Fuel Era.* Cambridge, MA: MIT Press, 2015.

Rabinbach, Anson. *The Human Motor: Energy, Fatigue, and the Origins of Modernity.* Berkeley: University of California Press, 1992.

Roberts, Paul. *The End of Oil: On the Edge of a Perilous New World.* New York: Houghton Mifflin, 2004.

Robertson, Thomas. *The Malthusian Moment: Global Population Growth and the Birth of American Environmentalism.* New Brunswick, NJ: Rutgers University Press, 2012.

Rodgers, Daniel T. *Age of Fracture.* Cambridge, MA: Belknap Press, 2012.

Roitman, Janet. *Anti-Crisis.* Durham, NC: Duke University Press, 2014.

Rome, Adam. *Bulldozer in the Countryside: Suburban Sprawl and the Rise of American Environmentalism.* New York: Cambridge University Press, 2001.

Rome, Adam. " 'Give Earth a Chance': The Environmental Movement and the Sixties." *Journal of American History* 90, no. 2 (September 2003): 525–554.

Sabin, Paul. *The Bet: Paul Ehrlich, Julian Simon, and Our Gamble over Earth's Future.* New Haven, CT: Yale University Press, 2013.

Sabin, Paul. *Crude Politics: The California Oil Market, 1900–1940.* Berkeley: University of California Press, 2005.

Sandalow, David. *Freedom from Oil: How the Next President Can End the United States' Oil Addiction*. New York: McGraw-Hill, 2008.

Sandbrook, Dominic. *Mad as Hell: The Crisis of the 1970s and the Rise of the Populist Right*. New York: Alfred A. Knopf, 2011.

Sargent, Daniel. *A Superpower Transformed: The Remaking of American Foreign Relations in the 1970s*. Oxford: Oxford University Press, 2015.

Schlosser, Kolson. "Malthus at Midcentury: Neo-Malthusianism as Bio-Political Governance." *Cultural Geographies* 16 (2009): 465–484.

Schneider-Mayerson, Matthew. *Peak Oil: Apocalyptic Environmentalism and Libertarian Political Culture*. Chicago: University of Chicago Press, 2015.

Schoijet, Mauricio. "Limits to Growth and the Rise of Catastrophism." *Environmental History* 4, no. 4 (October1999): 515–530.

Scott, Richard. *IEA: The First 20 Years, 1974–1994*. Vol. 1. Paris: OECD/IEA, 1994.

Seiler, Cotten. *Republic of Drivers: A Cultural History of Automobility in America*. Chicago: University of Chicago Press, 2008.

Self, Robert O. *All in the Family: The Realignment of American Democracy since the 1960s*. New York: Hill and Wang, 2012.

Self, Robert O. *American Babylon: Race and the Struggle for Postwar Oakland*. Princeton, NJ: Princeton University Press, 2003.

Shank, J. B. "Crisis: A Useful Category of Post-Social Scientific Historical Analysis?" *American Historical Review* 113, no. 4 (October 2008): 1090–1099.

Sharrett, Christopher. *Crisis Cinema: The Apocalyptic Idea in Postmodern Narrative Film*. Washington, DC: Maisonneuve Press, 1993.

Shi, David. *The Simple Life: Plain Living and High Thinking in American Culture*. Athens: University of Georgia Press, 1985.

Shiller, Robert J. "Narrative Economics." *American Economic Review* 107, no. 4 (2017): 967–1004.

Shiva, Vandana. *Soil Not Oil: Environmental Justice in a Time of Climate Crisis*. Cambridge: South End Press, 2008.

Siegel, Mike. "Passion and Poetry: Sam's Trucker Movie." Special Feature for *Convoy*. Directed by Sam Peckinpah, 1978. Los Angeles, CA: Twentieth Century–Fox Home Entertainment, 2015, DVD.

Slobodian, Quinn. *Globalists: The End of Empire and the Birth of Neoliberalism*. Cambridge, MA: Harvard University Press, 2018.

Sluyterman, Keetie. *Keeping Competitive in Turbulent Markets, 1973–2007: A History of Royal Dutch Shell*. Vol. 3. New York: Palgrave Macmillan, 2014.

Smil, Vaclav. *Energy and Civilization: A History*. Cambridge, MA: MIT Press, 2017.

Sodowsky, Alice, Roland Sodowsky, and Stephen Witte. "The Epic World of American Graffiti." *Journal of Popular Film* 4, no. 1 (1975): 47–55.

Starn, Randolph. "Historians and 'Crisis.'" *Past and Present* 52 (August 1971): 3–22.

Stein, Judith. *Pivotal Decade: How the United States Traded Factories for Finance in the Seventies*. New Haven, CT: Yale University Press, 2010.

Stewart, Janet. "Sociology, Culture and Energy: The Case of Wilhelm Ostwald's

'Sociological Energetics'—A Translation and Exposition of a Classic Text." *Cultural Sociology* 8, no. 3 (2014): 333–350.

Stoekl, Allan. *Bataille's Peak: Energy, Religion, and Postsustainability*. Minneapolis: University of Minnesota Press, 2007.

Swan, Edward J. *Building the Global Market: A 4000 Year History of Derivatives*. London: Kluwer Law International, 2000.

Szeman, Imre. *On Petrocultures: Globalization, Culture, and Energy*. Morgantown: West Virginia University Press, 2019.

Szeman, Imre, and Eric Cazdyn. *After Globalization*. New York: Wiley-Blackwell, 2011.

Szeman, Imre, and Caleb Wellum. "Energy Humanities." In *Oil Spaces: Exploring the Global Petroleumscape*, edited by Carola Hein, 211–225. New York: Routledge, 2021.

Tadiar, Neferti X. M. "Life-Times in Fate Playing." *South Atlantic Quarterly* 111, no. 4 (Fall 2012): 783–802.

Tertzakian, Peter. *A Thousand Barrels a Second: The Coming Oil Break Point and the Challenges Facing an Energy Dependent World*. New York: McGraw-Hill, 2006.

Trim, Henry. "A Quest for Permanence: The Ecological Visioneering of John Todd and the New Alchemy Institute." In *Groovy Science: Knowledge, Innovation, and American Counterculture*, edited by David Kaiser and Patrick McCray. Chicago: University of Chicago Press, 2016, 142–171.

Turner, James Morton. *The Republican Reversal: Conservatives and the Environment from Nixon to Trump*. Cambridge, MA: Harvard University Press, 2018.

Uhlman, James Todd, and John Heitmann. "Stealing Freedom: Auto Theft and Autonomous Individualism in American Film." *Journal of Popular Culture* 48, no. 1 (2015): 86–101.

US Commodity Futures Trading Commission. "CFTC History in the 1980s." http://www.cftc.gov/About/HistoryoftheCFTC/history_1980s.

Venn, Fiona. *The Oil Crisis*. London: Longman, 2002.

Venture, Patricia. *Neoliberal Culture: Living with American Neoliberalism*. New York: Routledge, 2012.

Walker, Jeremy, and Melinda Cooper. "Genealogies of Resilience: From Systems Ecology to the Political Economy of Crisis Adaptation. *Security Dialogue* 42, no. 2 (2011): 143–160.

Walker, J. Samuel. "The Nuclear Power Debate of the 1970s." In *American Energy Policy in the 1970s*, edited by Robert Lifset. Norman: University of Oklahoma Press, 2014.

Wapshott, Nicholas. *Keynes Hayek: The Clash that Defined Modern Economics*. New York: W. W. Norton, 2011.

Warde, Paul, Libby Robin, and Sverker Sörlin. *The Environment: A History of the Idea*. Baltimore: Johns Hopkins University Press, 2018.

Wells, Christopher W. *Car Country: An Environmental History*. Seattle: University of Washington Press, 2012.

Westad, Odd Arne. *The Global Cold War: Third World Interventions and the Making of Our Times*. New York: Cambridge University Press, 2007.

White, Richard. *The Organic Machine: The Remaking of the Columbia River*. New York: Hill and Wang, 1996.

Williams, Jeffrey C. *The Economic Function of Futures Markets*. New York: Cambridge University Press, 1986.

Williams, Jeffrey C. "The Origins of Futures Markets." *Agricultural History* 56, no. 1 (January 1982): 306–316.

Williams, Raymond. "Culture Is Ordinary." In *Raymond Williams on Culture and Society: Essential Writings*, edited by Jim McGuigan. Washington, DC: Sage, 2014.

Williams, Raymond. *The Long Revolution*. Westport, CT: Greenwood Press, 1961.

Williams, Raymond. *Marxism and Literature*. Oxford University Press, 1977.

Wohlstetter, John. *Herman Kahn: Public Nuclear Strategy 50 Years Later: A Compendium of Highlights from Herman Kahn's Works on Nuclear Strategy*. Washington, DC: Hudson Institute, September 2010.

Yamazato, Katsunori. "How to Be in This Crisis: Gary Snyder's Cross-Cultural Vision in *Turtle Island. Critical Essays on Gary Snyder*. Edited by Patrick D. Murphy, 230–231. Boston: G. L. Hal, 1991.

Yaqub, Salim. *Imperfect Strangers: Americans, Arabs, and U.S.–Middle East Relations in the 1970s*. Ithaca, NY: Cornell University Press, 2016.

Yergin, Daniel. *The Prize: The Epic Quest for Oil, Money, and Power*. 2nd ed. New York: Free Press, 2009.

Zaloom, Caitlin. *Out of the Pits: Traders and Technology from Chicago to London*. Chicago: University of Chicago Press, 2006.

Zaretsky, Natasha. *No Direction Home: The American Family and the Fear of National Decline, 1968–1980*. Chapel Hill: University of North Carolina Press, 2007.

Zelko, Frank. *Make It a Green Peace! The Rise of Countercultural Environmentalism*. New York: Oxford University Press, 2013.

Zuckerman, Gregory. *The Frackers: The Outrageous Story of the New Billionaire Wildcatters*. New York: Portfolio, 2013.

Index

Page numbers in *italic* refer to illustrations.